BEFORE THIS DECADE IS OUT

Personal Reflections on the Apollo Program

Edited by GLEN E. SWANSON

Introduction to the Dover Edition by PAUL DICKSON

DOVER PUBLICATIONS, INC.
Mineola, New York

Bibliographical Note

This Dover edition, first published in 2012, is an unabridged republication of the work originally published in the NASA History Series as NASA SP-4223, Washington, D.C., in 1999. A new Introduction by Paul Dickson has been added to this edition.

International Standard Book Number
ISBN-13: 978-0-486-48383-2
ISBN-10: 0-486-48383-5

Manufactured in the United States by Courier Corporation
48383502
www.doverpublications.com

Introduction
to the Dover Edition

While the vast majority of books in the NASA History Series are tight, well-written, and highly vetted narratives, *Before This Decade Is Out: Personal Reflections on the Apollo Program* is none of these. In contrast, it is a loose, plain spoken history of the Apollo Space Program as told through the stories of fourteen diverse individuals who were part of it. Johnson Space Center Historian Glen E. Swanson relied on the large collection of oral histories available within the agency to compile these recollections.

The most historically significant insights are revealed in the interviews with those involved in the administration operation of Apollo such as rocket developer Wernher von Braun, Administrator James E. Webb, and astronaut Charles M. Duke, Jr. Important decisions, tense moments, and squabbles between key players are discussed. To a person, the decision-makers and front-line actors are direct and honest. Few, if any, punches are pulled.

In one example, Eugene F. Kranz, flight director, captures the spirit of NASA in the wake of the Apollo 1 fire that claimed the lives of three astronauts. He talks about the agency becoming complacent, paying the price for that and then rededicating itself to the lunar mission with "pure, raw leadership" by "incredibly talented and capable people." He also dramatically describes the tension in the control room when the Apollo 11 Lander touched down with only 15 seconds of fuel left in the tank.

There are also interviews that are instructional such as the one with scientist-astronaut Harrison H. Schmitt, which amounts to an impromptu scientific history of the moon, as well as a discussion of several options for lunar solutions to energy issues. One of these is to employ the lunar surface as large collector of solar energy which could then be transmitted back to earth. He also points out that a metric ton of Helium 3 contains as much energy as $2 billion dollars worth of coal burned today as a source of electricity. That amount of coal provides the electricity for 10 million Americans for about a year. "As an old economic geologist," said Schmidt, "when I hear about a resource that may be worth two billion dollars a metric ton, I can start to think of all sorts of ways that I might get access to that resource."

Though some of the most prominent people in the history of the program are here so are some whose roles have tended to get buried deep in the footnotes of the grand story. Geneva B. Barnes, for instance, is included for her role in running the world tour of the Apollo 11 astronauts following the first lunar landing. It is a fascinating behind the scenes account of a moment of glory which was also a logistical nightmare. For example, one problem unique to this tour was that when members of the tour became ill, Barnes had to deal with rumors that the astronauts had brought back a lunar sickness and that all the staff were becoming ill as a result of being exposed to them. When the tour got to London, the NASA doctor had to go on national television to deny that any member of the staff was moon sick. Barnes is remarkably honest in reporting her own gaffes—such as when she stepped out of her hotel room barefoot and in curlers only to be greeted by the Prime Minister of Australia.

Another example of the eclectic nature of the interview subjects is James H. "Harry" Guin, director of Propulsion Test Operations for the Stennis Space Center in Mississippi, who, among other things, discusses the early days of the facility when it was an open animal range where wild pigs and piglets had to periodically be rounded up, which was no simple feat. Then there were the huge number of snakes, many of which were poisonous, that workers at the facility had to contend with, plus the mosquitoes.

How bad were the mosquitoes? For starters, in the early days cattle were killed because the swarming mosquitoes would be ingested during breathing and suffocate them. Crews building fences around the base actually went on strike because the mosquitoes were so bad and no existing repellent could stop them from being badly bitten. A solution was finally found involving the "absolute cheapest, smelliest perfume" that could be found.

How, one might ask, did NASA pick this spot? That's answered in an interview with Arthur J. "Jack" Rogers, Jr., chief of the Facilities Engineering Office, also at the Stennis Space Center, who acknowledges that the device employed for this particular mission was a Texaco road map which he and his fellow decision makers broke out in an attempt to find a location near navigable waterways, close to the Michoud Assembly Facility, on high ground and that was sparsely populated—the pigs, snakes, and mosquitos were a bonus.

All and all, *Before This Decade Is Out* is a very readable and totally engaging book.

PAUL DICKSON

BEFORE THIS DECADE IS OUT

Personal Reflections
on the Apollo Program

CONTENTS

FOREWORD

As the 20th century ends and we approach the beginning of a new millennium, future generations will look back and recall the leaving of our home planet as a defining moment in the history of humankind.

The story of Apollo has been told and retold many times by those fortunate enough to have experienced space travel first hand—the astronauts themselves. Those who managed to leave the cradle of the Earth and walk upon the surface of the Moon offer a unique perspective shared firsthand by only a handful of people chosen to represent humankind in the culmination of this great adventure. As we reflect on the 30th anniversary of Apollo's first landing, we should not overlook other accounts of this monumental achievement as offered by those representing the half a million people associated with the program. During the 1960s, these people joined in a national effort to bring to completion President John F. Kennedy's vision of landing Americans on the Moon and returning them safely to Earth before the end of the decade.

For a project as massive as the Apollo program, history may distance itself to the extent where modern interpretation distills a feeling that such events took place without extensive human involvement. Nothing could be further removed from the truth. Through the verbal accounts offered by the oral histories such as presented in this volume, we are reintroduced to the critical human factor which is the essence of any history. People made Apollo happen and it is important to preserve their thoughts, feelings, and recollections for future generations. The oral histories presented in this volume offer a sample of what NASA has done to preserve the story of Apollo as part of our nation's human spaceflight heritage.

The accounts included in this book are a small sampling of the large number of oral histories that have been conducted under the auspices of the NASA history program, since near the beginning of the Agency. They also represent the many personal contributions made during Project Apollo, the single largest peacetime endeavor in American history. These recollections span the origins, management, and completion of that enormous effort and measurably enhance our appreciation of its difficulty. I am pleased that the comments of some of the key individuals involved in Project Apollo are being preserved by NASA and made available through this book.

The people who are quoted in this book were among the top leaders of NASA. All of them played a prominent part in the conduct and accomplishments of Apollo. As one of those who knew and watched these individuals lead, I have a particular sense of their statements. I always had the feeling of having been granted a special privilege to participate and work on the Mercury, Gemini, and Apollo programs. The contents of this book reveal that these people had similar experiences. They all recognized that it took literally thousands of dedicated people to bring these efforts to fruition and that it was up to them to provide the necessary leadership to allow all of the workers on the project to accomplish their tasks. It was a wondrous thing to watch. Anyone interested in the underlying strength of NASA in this time period will find these accounts a fascinating read.

Christopher C. Kraft, Jr.
Director, Flight Operations,
    NASA Johnson Space Center, 1963–1969
Director,
    NASA Johnson Space Center, 1972–1982

ACKNOWLEDGMENTS

This volume takes advantage of the large collections of oral histories available within NASA which have been conducted and gathered since the early beginnings of the agency. More recent efforts include the Johnson Space Center Oral History Project (JSC OHP) which was established in 1996, under the direction of JSC Director George Abbey, to document and record the memories of those individuals involved in the history of human space flight. We pursued the task of creating this book when the opportunity presented itself to utilize many of the oral histories resulting from this and other oral history projects, specifically focusing on those people who had a role in the successful completion of Project Apollo in order to publish their personal reflections in commemoration of the 30th anniversary of Apollo 11. In doing so, we incurred numerous debts.

A great deal of time and effort went into researching, interviewing, transcribing, and duplicating each of the oral histories present in this volume. For those interviews conducted and used in this volume as part of the JSC OHP, a special note of thanks is extended to Bill Larsen, Carol Butler, Summer Chick Bergen, Rebecca Wright, Michelle Kelly, Kevin Rusnak, Tim Farrell, and Sasha Adams Tarrant. In addition, a special note of thanks to my JSC Branch Manager, Peggy Wooten, who was patient in allowing me time to do research.

Numerous past oral histories were also researched and utilized. Those that helped in locating these include Janet Kovacevich of the JSC Science and Technical Information Center. Joan Ferry and Lois Morris of the JSC Collection at Rice University were invaluable in researching, locating, and making copies of hundreds of pages of interview transcripts for review. Early interviews used in this work include those conducted by H. George Frederickson, Henry J. Anna, Barry Kelmachter, Roger E. Bilstein, Robert Sherrod, John Beltz, Eugene M. Emme, Jay Holmes, Addison Roghrock, Ivan Ertel, and James M. Grimwood.

Among the many previous oral history projects conducted by and through NASA, one stands out as being the most complete in both size and detail. Much gratitude is given to the late Robert Sherrod who, from 1968-1974, interviewed dozens of key players in the Apollo program. A career journalist, he covered General MacArthur's campaign against the Japanese in the South Pacific during WWII. After the war, he was one of *Time-Life's* leading correspondents, specializing in defense matters as well as covering the early years of the U.S. manned space program. He was the final editor of the *Saturday Evening Post* before its demise. *Time-Life* contracted Sherrod to research and write a detailed history of the Apollo program but sadly enough, the book never materialized. In its wake, he left to NASA hundreds of pages of detailed transcripts meticulously annotated for future researchers, several of which were used for this book.

For those interviews that survive only in audio tape form, recovery and restoration work was needed during the course of research. JSC OHP technical support members Paul Rollins and Franklin Tarazona proved invaluable in quickly responding to various duplication requests and technical questions.

The numerous photos that appear throughout the text were gleaned from several sources, including the editor's own personal collection. Mike Gentry of the JSC Media Resource Center came through with several last minute photo requests. Mary Wilkerson and Linda Turney of the JSC Still Image Repository also spent time in responding to numerous photo queries. Many photos also came to be used in this book as a result of Brian Nicklas, of the Smithsonian's National Air and Space Museum, who led this editor to their research office's newly acquired Herb Desind Collection.

Much of this text would not have been possible initially without the willingness of those individuals (both past and present) featured in these chapters to take the time and effort to share their personal reflections. A special thank you is extended to James E. Webb, Thomas O. Paine, Wernher von Braun, Robert R. Gilruth, George E. Mueller, Eugene F. Kranz, Arthur J. Rogers, Jr., James H. "Harry" Guin, Glynn S. Lunney, Geneva B. Barnes, Charles M. Duke, Jr., Harrison H. "Jack" Schmitt, George M. Low, and Maxime A. Faget.

Numerous people at other NASA Centers associated with historical study, technical information, and the mechanics of publishing helped in myriad ways in the preparation of this documentary history. Virginia Butler of the NASA Stennis Space Center provided several oral histories that were used as part of the Stennis Space Center's History Project, done in conjunction with the Mississippi Oral History Program of the University of Southern Mississippi. At NASA Headquarters, Stephen J. Garber helped in the final proofing. M. Louise Alstork and Nadine J. Andreassen of the NASA History Office performed editorial and proofreading work on the project; and the staffs of the NASA Headquarters Library, the Scientific and Technical Information Program, and the NASA Document Services Center provided assistance in locating and preparing for publication the documentary materials used in this work. The NASA Headquarters Printing and Design Office developed the layout and handled printing. Specifically, we wish to acknowledge the work of Jane E. Penn, Chris Pysz, Geoff Hartman, Kelly L. Rindfusz, and Joel Vendette for their editorial and design work. In addition, Michael Crnkovic, Stanley Artis, and Jeffrey Thompson saw the book through the publication process. Thanks are due them all.

Glen E. Swanson
Historian
NASA Johnson Space Center
July 20, 1999

INTRODUCTION
THE LEGACY OF PROJECT APOLLO
By Roger D. Launius

July 1999 marks the 30th anniversary of the epochal lunar landing of *Apollo 11* in the summer of 1969. Although President John F. Kennedy had made a public commitment on May 25, 1961, to land an American on the Moon by the end of the decade, up until that time Apollo had been all promise. Now the realization was about to begin. Kennedy's decision had involved much study and review prior to making it public, and his commitment had captured the American imagination, generating overwhelming support. Project Apollo had originated as an effort to deal with an unsatisfactory situation (world perception of Soviet leadership in space and technology), and it addressed these problems very well. Even though Kennedy's political objectives were achieved essentially with the decision to go to the Moon, over the years Project Apollo took on a life of its own over the years and left an important legacy to both the nation and the proponents of space exploration. Its success was enormously significant, coming at a time when American society was in crisis.

A unique confluence of political necessity, personal commitment, activism, scientific and technological ability, economic prosperity, and public mood made possible the 1961 decision to carry out an aggressive lunar landing program. It then fell to NASA, other organizations of the Federal Government, and the aerospace community to accomplish the task set out in a few short paragraphs by the president. By the time that the goal was accomplished in 1969, only a few of the key figures associated with the decision were still in leadership positions in the government. Kennedy fell victim to an assassin's bullet in 1963, and science adviser Jerome B. Wiesner returned to MIT soon afterwards. Lyndon B. Johnson, of course, succeeded Kennedy as president but left office in January 1969 just a few months before the first landing. NASA Administrator James E. Webb resolutely guided NASA through most of the 1960s, but his image was tarnished by, among other things, a 1967 Apollo accident that killed three astronauts. Consequently, he retired from office under something of a cloud in October 1968. Several other early supporters of Apollo in Congress and elsewhere died during the 1960s and never saw the program successfully completed.

The first Apollo mission to the Moon was the flight of *Apollo 8*. On December 21, 1968, the first manned spacecraft to leave the gravitional well of the Earth was launched atop a *Saturn V* booster from the Kennedy Space Center. Three astronauts were aboard—Frank Borman, James A. Lovell, Jr., and William A. Anders—for a historic mission to orbit the Moon. At first, that mission had been planned as a flight to test Apollo hardware in the relatively safe confines of low Earth orbit, but Apollo Program Manager George M. Low, of the Manned Spacecraft Center at Houston, Texas, and Samuel C. Phillips, Apollo Program Manager at NASA Headquarters, obtained approval to make it a circumlunar flight. The advantages of this mission were important, both in technical and scientific knowledge gained as well as in a public demonstration of what the U.S. could achieve.

After *Apollo 8* made one and a half Earth orbits, its third stage began a burn to put the spacecraft on a lunar trajectory. It orbited the Moon on December 24 and 25 and then fired the boosters for a return flight. It "splashed down" in the Pacific Ocean on December 27. That flight was such an enormously significant accomplishment because it came at a time when American society was in crisis over Vietnam, race relations, urban problems, and a host of other difficulties. If only for a few moments the Nation united as one to focus on this epochal event. Two more Apollo missions occurred before the climax of the program, testing critical systems and procedures and confirming that the time had come for a lunar landing.

That landing came during the flight of *Apollo 11*, which lifted off on July 16, 1969, and, after confirmation that the hardware was working well, began the three-day trip to the Moon. Then, at 4:18 p.m. EST on July 20, 1969, the Lunar Module—with astronauts Neil A. Armstrong and Edwin E. "Buzz" Aldrin aboard—landed on the lunar surface while Michael Collins orbited overhead in the Apollo command module. After checkout, Armstrong set foot on the surface, telling millions who saw and heard him on Earth that it was "one small step for [a] man—one giant leap for mankind." Aldrin soon followed him out and the two plodded around the landing site in the one-sixth lunar gravity, planted an American flag but omitted claiming the land for the U.S. as had been routinely done during European exploration of the Americas, collected soil and rock samples, and set up scientific experiments. The next day they launched back to the Apollo capsule orbiting overhead and began the return trip to Earth, splashing down in the Pacific on July 24.

This flight rekindled the excitement felt in the early 1960s during the first Mercury flights, and set the stage for later Apollo landing missions. An ecstatic reaction enveloped the globe, as everyone shared in the success of the mission. Ticker tape parades, speaking engagements, public relations events, and a world tour by the astronauts served to create good will both in the U.S. and abroad. Five more landing missions followed at approximately six-month intervals through December 1972, each of them increasing the time spent on the Moon. The scientific experiments placed on the Moon and the lunar soil samples returned have provided grist for scientists' investigations ever since. The scientific return was significant, but the program did not answer conclusively the age-old questions of lunar origins and evolution. Three of the latter Apollo missions also used a lunar rover vehicle to travel in the vicinity of the landing site, effectively extending the range that the astronauts coul safely explore, but none of them equaled the initial excitement of Apollo 11.

Project Apollo in general, and the flight of *Apollo 11* in particular, should be viewed as a watershed in the Nation's history. It was an endeavor that demonstrated both the technological and economic virtuosity of the United States and established national preeminence over rival nations—the primary goal of the program when first envisioned by the Kennedy administration in 1961. It had been an enormous undertaking, costing $25.4 billion (about $95 billion in 1999 dollars) with only the building of the Panama Canal rivaling the Apollo program's size as the largest nonmilitary technological endeavor ever undertaken by the United States and only the Manhattan Project being comparable in a wartime setting.

There are several important legacies (or conclusions) about Project Apollo that need to be remembered at the thirtieth anniversary of the *Apollo 11* landing. First, and probably most important, the Apollo program was successful in accomplishing the political goals for which it had been created. Kennedy had been dealing with a Cold War crisis in 1961 brought on by several separate factors—the Soviet orbiting of Yuri Gagarin and the disastrous Bay of Pigs invasion only two of them—that Apollo was designed to combat. At the time of the *Apollo 11* landing, Mission Control in Houston flashed the words of President Kennedy announcing the Apollo commitment on its big screen. Those phrases were followed with these: "TASK ACCOMPLISHED, July 1969." No greater understatement could probably have been made. Any assessment of Apollo that does not recognize the accomplishment of landing an American on the Moon and safely returning before the end of the 1960s is incomplete and inaccurate, for that was the primary goal of the undertaking.

Second, Project Apollo was a triumph of management in meeting the enormously difficult systems engineering and technological integration requirements. James E. Webb, the NASA administrator at the height of the program between 1961 and 1968, always contended that Apollo was much more a management exercise than anything else, and that the technological challenge, while sophisticated and impressive, was also within grasp. More difficult was ensuring that those technological skills were properly managed and used. Webb's contention was confirmed by the success of Apollo. NASA leaders had to acquire and organize unprecedented resources to accomplish the task at hand. From both a political and technological perspective, management was critical. For seven years after Kennedy's Apollo decision, through October 1968, James Webb politicked, coaxed, cajoled, and maneuvered for NASA in Washington. In the process, he acquired for the agency sufficient resources to meet Apollo requirements.

More to the point, NASA personnel employed a "program

management" concept that centralized authority over design, engineering, procurement, testing, construction, manufacturing, spare parts, logistics, training, and operations. The management of the program was recognized as critical to Apollo's success in November 1968, when *Science* magazine, the publication of the American Association for the Advancement of Science, observed:

> *In terms of numbers of dollars or of men, NASA has not been our largest national undertaking, but in terms of complexity, rate of growth, and technological sophistication it has been unique.... It may turn out that [the space program's] most valuable spin-off of all will be human rather than technological: better knowledge of how to plan, coordinate, and monitor the multitudinous and varied activities of the organizations required to accomplish great social undertakings.*

Understanding the management of complex structures for the successful completion of a multifarious task was a critical outgrowth of the Apollo effort.

Third, Project Apollo forced the people of the world to view the planet Earth in a new way. *Apollo 8* was critical to this change, for on its outward voyage, the crew focused a portable television camera on Earth and for the first time humanity saw its home from afar, a tiny, lovely, and fragile "blue marble" hanging in the blackness of space.

When the Apollo 8 spacecraft arrived at the Moon on Christmas Eve of 1968 the image of Earth was even more strongly reinforced when the crew sent images of the planet back while reading from the Book of Genesis, "And God created the heavens and the Earth, and the Earth was without form and void"—before sending holiday greetings to humanity. Writer Archibald MacLeish summed up the feelings of many people when he wrote at the time of Apollo that "To see the Earth as it truly is, small and blue and beautiful in that eternal silence where it floats, is to see ourselves as riders on the Earth together, brothers on that bright loveliness in the eternal cold—brothers who know now that they are truly brothers." The modern environmental movement was galvanized in part by this new perception of the planet and the need to protect it and the life that it supports.

Finally, the Apollo program, while an enormous achievement, left a divided legacy for NASA and the aerospace community. The perceived "golden age" of Apollo created for the agency an expectation that the direction of any major space goal from the president would always bring NASA a broad consensus of support and provide it with the resources and license to dispense them as it saw fit. Something most NASA officials did not understand at the time of the Moon landing in 1969, however, was that Apollo had not been a normal situation and would not be repeated. The Apollo decision was, therefore, an anomaly in the national decision-making process.

The dilemma of the "golden age" of Apollo has been difficult to overcome, but moving beyond the Apollo program to embrace future opportunities has been an important goal of the agency's leadership in the recent past. Exploration of the solar system and the universe remains as enticing a goal and as important an objective for humanity as it ever has been. Project Apollo was an important early step in that ongoing process of exploration.

NASA Administrator James E. Webb seated in the Gemini rendezvous and docking simulator during an August 7, 1965 visit to the Manned Spacecraft Center (renamed the Johnson Space Center in 1973). (NASA Photo S-65-28481.)

CHAPTER 1

JAMES E. WEBB

(1 9 0 6 - 1 9 9 2)

James Edwin Webb was the second administrator of the National
Aeronautics and Space Administration, serving between 1961 and
1968. During his tenure, NASA developed the modern techniques necessary to coordinate and direct the most unique and complex
technological enterprise in human history, the sending of human beings
to the Moon and bringing them safely back to the Earth.

Born on October 7, 1906, in Granville County, North Carolina, he
was the son of John Frederick and Sarah Gorham. Webb was educated
at the University of North Carolina, where he received an A.B. in
Education 1928. He also studied law at George Washington University
and was admitted to the bar of the District of Columbia in 1936.

Webb enjoyed a long career in public service, coming to Washington
in 1932 and serving as secretary to Congressman Edward W. Pou, 4th
North Carolina District, chairman of House Rules Committee, until
1934. He then served as assistant in the office of O. Max Gardner,
attorney and former governor of North Carolina, in Washington, D.C.,
between 1934 and 1936. In 1936, Webb became secretary-treasurer
and later vice president of the Sperry Gyroscope Company in Brooklyn,
New York, before entering the U.S. Marine Corps in 1944. After
World War II, Webb returned to Washington and served as executive

1

assistant to O. Max Gardner, by then undersecretary of the Treasury, before being named as director of the Bureau of the Budget in the Executive Office of the President, a position he held until 1949. President Harry S. Truman then asked Webb to serve as undersecretary of state. When the Truman administration ended early in 1953, Webb left Washington for a position in the Kerr-McGee Oil Corp. in Oklahoma. Webb's long experience in Washington would pay handsomely during his years at NASA, where he pressed for Federal support for the space program and dealt with competing interests on Capitol Hill and in the White House.

James Webb returned to Washington on February 14, 1961, when he accepted the position of administrator of NASA. For seven years after Kennedy's 1961 lunar landing announcement, through October 1968, James Webb politicked, coaxed, cajoled, and maneuvered for NASA in Washington. The longtime Washington insider proved a master at bureaucratic politics. In the end, through a variety of methods, Administrator Webb built a seamless web of political liaisons that brought continued support for and resources to accomplish the Apollo Moon landing on the schedule President Kennedy had announced. He left NASA in October 1968, just as Apollo was nearing a successful completion.

After retiring from NASA, Webb remained in Washington, D.C., serving on several advisory boards, including as a regent of the Smithsonian Institution. He died on March 27, 1992, in Washington, D.C.

Editor's note: The following are edited excerpts from an original interview with James E. Webb, conducted on May 15, 1969, in Washington, D.C., by H. George Frederickson, Henry J. Anna, and Barry Kelmachter. Original interview available in the NASA Historical Reference Collection, NASA History Office, Washington, D.C.

Could you give us some idea of how the concept of project management [for Project Apollo] was developed within NASA and how they implemented it?

. . . In a way my own experience added a different element to that of Dr. [Hugh L.] Dryden [NASA deputy administrator] and of Dr. [Robert C.] Seamans [NASA associate administrator]. The three of us together had a better knowledge than most of how you could get large projects successfully accomplished. We had a clear understanding that you weren't going to get the end result with a contract—you had to solve the technical problems. We knew we had to have the contracting relationships with industry and the research grant relationships with university scientists and engineers, but we also had to have the in-house technical capability to monitor those contracts and grants and to provide the leadership dynamics to modify all or any part of these factors whenever feedback showed they should be modified. When experts were saying the job was too big for the country, that there were not enough scientists and engineers to do it, we had a sufficient knowledge of the capabilities of the country to know the job could be done if we effectively used these capabilities.

. . . We wanted industry to learn to do research and development under conditions where there would be no large production contracts to follow. This was a major educational job with respect to American industry because the trend had been to come in with a beautiful brochure, obtain the R&D contract, and then work out any problems that developed with little regard for efficiency, knowing that a production contract would follow on the current or another system and they'd make up any losses. In NASA, we weren't going to have the money to make up any losses because we weren't going to have any large production contracts, so our contractors had to learn to do R&D and to accept government men with knowledge as close collaborators in a new pattern for developing and applying technical solutions to problems.

We spoke with project managers and asked what was most critical in their day-to-day work and we got a pretty uniform judgment. With a project like Apollo, a highly mandated program, they said schedule. Their response was "I've got to meet my schedule with as much performance as I can get to that schedule, but I've got to meet that schedule." Regarding cost, there response, in general was that: "I feel I've been well enough funded to do virtually what I have to do." Do you concur with that kind of general attitude?

Not completely. This was an attitude that we fostered from the top, but we always kept a foot handy to the brake and a knowledge of when it might be needed, and we wouldn't let that get out of hand. It was not out of control. There are many cases where there is a tendency to do this in a mechanical way and in NASA they were stopped cold by top management who said you've got to do it a different way.

So as long as they [the contractors] are meeting schedules with adequate performance, NASA would continue to support them?

No, it is not a question of just adequate performance. My attitude was if you have 200 project managers who were adequate, you should still remove 10 percent of them every year to keep the pressure on them to do better.

From your standpoint, what were your toughest problems? Did you worry most about the technical problems or the administrative ones?

Personally I didn't because we had other people to make the critical judgments on technical things. I kept an overview of how I thought the various units were doing with respect to using good technical judgment. I knew that when Dr. Seamans went out to the centers and contractors to personally look at the hardware and reach a technical as well as overall judgment, that his judgment would feed into the circle of Dryden, Seamans, and myself. I knew that when people were talking to Dryden about technical matters of grave concern, he would inform Seamans and me. I knew that when industry executives came to tell me of their concerns, I would pass these on to Dryden and Seamans, as well as others who had a need to know. I guess my way of thinking about it was related more to which matter is causing the gravest concern, rather than what is the precise technical margin required for success. We were risk takers and had to be to get the job done.

Relating to project management, we frequently found project managers talking about their relations with the Center Directorate and came away with the feeling that there are important differences between Houston [home of the Manned Spacecraft Center (MSC), renamed the Johnson Space Center in 1973] and Huntsville [home of the Marshall Space Flight Center (MSFC)]. Is this true?

Sure. And there are important differences within Huntsville. The care and feeding of Huntsville as an organization capable of spending a billion, eight hundred million dollars a year, with a large part of it going outside, are very different than a project manager who wants to get one stage of one booster done. The question of how Huntsville wants to approach the building of a family of boosters and how Houston wants to approach the flying of a family of manned spacecraft are two different things.

Would you characterize it any further than that?

Yes, I'd say that in an organization like NASA, there's a certain element of lack of consideration of the total problem on the part of every specialist. At one point, some felt that Wernher von Braun [director of MSFC] and his people didn't fully understand the pilot-confidence element of flying men in our first high performance, marginally adequate, rocket-propelled craft like the X-15 or the Mercury spacecraft. [Robert L.] Gilruth [director of the

NASA Administrator James E. Webb, second from right, touring Marshall Space Flight Center with President John F. Kennedy, Vice President Lyndon Johnson, and other guests on September 11, 1962. (NASA Photo, available on NASA Image Exchange: *http://www.nix.nasa.gov.*)

Manned Spacecraft Center] and others tended to feel that unmanned booster concepts shouldn't govern, that the requirements contributing to pilot confidence should be the paramount object of attention. These are natural human problems that arise and you get better results with a certain amount of competition and fighting for these different points of view. I wouldn't have wanted to change that. Many good results came from this. Where did we get Eberhard Rees [one of Wernher von Braun's chief lieutenants at MSFC] to send out to North American [an aerospace corporation located in southern California] to ride herd on the Block 2 Apollo? From Huntsville, not Houston.

The job of headquarters is to kind of umpire this thing and monitor it and control it?

I don't like the word umpire. I think positive leadership is more important. To keep it under reasonable control but also not squelch the strong individualistic tendencies of these people and to do everything necessary to make them want to get the work done.

Was there ever a feeling that in the area of administration the Manned Spacecraft Center in Houston was a little weak?

Oh yes, it [Houston] had to start from scratch and it was very weak in the beginning. Dryden, Seamans, and I created a group of people that could go out to the various manned space installations and, because these were very senior people from outside the government, they could examine what was going on at Houston and then say to [Robert] Gilruth and the others at MSC, "Look, here's a better way to do it." The Houston officials would take these suggestions from an outside group that they respected, a peer group, when they wouldn't take it as an order from headquarters.

The mandate idea [President Kennedy's mandate of "landing a man on the Moon by the end of the decade"] had an incredible effect on the [Apollo] program. Do you think that the concept of project management can be equally successful when it is not as heavily mandated?

Bear in mind that the public image that we had to land on the Moon in this decade was not the real management concept of NASA. This was sort of a political forward thrust, looking at the fact that maybe the Russians would do something big by 1967 at their 50th anniversary. What we had in mind was to try and build all the elements of a total space competence . . . We realized that man had developed an engine big enough for him to leave the Earth and move around in space, and that man's new ability to use the forces of gravity and inertia in ways that broke all the limits of the past could be decisive in international competition . . .

The lunar project to us was little more than a realistic requirement for space competence. To do Apollo and in the doing of it acquire a complete space competence . . . we realized there must be a continuing advance of scientific knowledge . . . and when the politicians, including the President, tended to say, "Well, gee, we've got a tight budget here, just concentrate on getting this Moon thing," we always said "No, our objective must be broader."

The idea that the lunar landing was a mandate is a sort of foolish concept shared by a lot of people in NASA. You notice I never used it in my congressional testimony. I always stated that we were developing the full range of space capabilities and demonstrating this with the lunar landing. It is an interesting fact that the lunar landing represented a requirement that we learn to use 98 percent of the energy it takes to keep going to Mars or Venus—that it's about the same requirement as to get the same payload to synchronous orbit. There were many reasons associated with the lunar concept. The public image that it was a mandate, the way our public relations people dealt with it as a commitment on the part of Congress, you never saw me use. So you simply can't say that the idea or concept of a "mandate" produced the result. It was NASA's drive to do the total job of developing a full capability in space. If we simply said that we're going to the Moon and had not initiated the scientific and technical work forward of that which was required for the period after the landing, we never would have gotten to the Moon. We had to draw on knowledge that came from work generally considered to be beyond the original Apollo concept in order to do Apollo.

To the leadership of a large multifaceted organization operating within a framework of law and precedent, the mandate idea has its negative side. When President [John F.] Kennedy made his first visit to Huntsville and after looking around, said to Wernher, "Well what do you need? What can I do to help you get your work done?" Wernher replied, "Give us the money, and give us complete freedom to spend it any way we want to." I immediately broke in, because this is a foolish idea in a country like this with its pattern of legislative controls on the use of appropriated funds. Nobody could do that. But you see, this was von Braun's idea of the mandate; give him complete freedom.

Would it be better for NASA not to look for another mandate like this, but to secure the commitment and consent from Congress and from the president to continue to develop this total capability in spaceflight activity?

President John F. Kennedy and NASA Administrator James T. Webb at the Launch Operations Complex (later renamed the Kennedy Space Center) during a tour of NASA facilities on November, 1963. (NASA Photo, available on NASA Image Exchange: *http://www.nix.nasa.gov*.)

I think it's utterly foolish to ask a political leader to commit himself for some 10 to 20 years in advance to a goal that's going to involve a lot of money and that can easily be attacked but is hard to defend. What you have to generally do is proceed as rapidly as you can toward things that you can gain acceptance of and always be prepared to change, to slow up for large obstacles or to thrust forward if that seems possible. Very few people realized that the Gemini program permitted us to stop with the completion of the flights and never go to Apollo, yet still have a lot of experience as a base for future thrusts. Of course, some goal or goals are necessary, but beyond Apollo, goals like a manned landing on Mars have such long lead-times that it is doubtful a president and a Congress will stake their full power on getting what you would call a mandate.

Theoretically then, we could stop at Apollo or we could continue a different or wider program?

It's the development, exploration and use of space that you are after, not a mandate to go to Mars . . . I think that in many ways Kennedy said we were going to the moon in this decade not on the basis of what NASA had told him, but on the basis of how he felt as President he could bid for support. That has little to do with project management, except that project management finds it hard to understand that their job is not what the President said it is. It's very hard for them to understand that, because they're generally people with blinders on to limit side vision in getting at their projects. That's what you want them to do. You don't want them to try to run NASA and run their project at the same time.

Do you think the general run-of-the-mill Congressman knows more about Apollo and its activities than he knows about other programs of the same magnitude? Has Apollo done a better job of informing Congress?

. . . *Life* magazine asked NASA to put on a week's demonstration for the heads of 20 companies that paid a billion dollars in taxes and had no space contracts. At the end of that program, the head of one of the biggest financial institutions in the United States got up and made a little speech. He said: "When I started out from New York a week ago, if anybody would have asked me if I was a supporter of this program, I would have said 'no'. Now a week later I've learned an awful lot and I can't say I'm 100 percent for it, but I'm about 95 precent for what I've seen." And then he pointed a finger at me and said, "What do you mean by doing all this without letting me know? . . ."

This was true of a lot of the members of Congress. The essence of Apollo's support that enabled us to make sure we had the votes when we needed them lay in the dedication of a few congressional leaders who were determined that the United States' future, its ability to hold and wield power and not become second-best in advanced technology, would not be put in jeopardy by failure to develop this new space medium. And there were a few who were determined to destroy the program. So there was this general contest among very powerful people which was decided in annual increments, by counting votes in the committees and in the House and Senate. There were a few of us who could talk to both sides and knew how to anticipate the votes . . . And because of this we were able to make a very simple assumption: "We have a strong base of support, so we're going to put in the program enough power to get airborne and enough power to get into orbital speed on this program . . ."

We put Nova [a project to develop a massive expendable booster larger than the Saturn V to enable direct ascent to the lunar surface] in the Apollo program as well as Saturn. Later, we found that we didn't need Nova, so we readjusted the funds without losing our base of support . . . Our statements to Congress made it clear that we had to make changes on a learn-as-you-go basis. This was accepted by the budget director, by the president and by leaders in Congress. What the general run-of-the-mill congressman thought or how he answered his people, depended a lot on his own situation. Many did not want to make a public commitment, but I've never known one where you went into his district with an astronaut . . . and, at the end of the day, after they had seen the tremendous response that this program drew from the younger people in the district, they would say, "This is fine, I'm for it."

I've noticed a lot of personnel shifts, from Headquarters out to various Centers, from the Centers back to Headquarters . . . Is this intentional?

Yes, a great deal of this was intentional . . . I made it clear to do all we could to encourage people to cooperate in these moves. We couldn't order people to move like you can in the military; we had to persuade them. Eventually, we got enough examples of people who had gained substantially personally from this that others then sought it.

One of the most interesting things [that happened to me] is when I failed to get a deputy director after Seamans left. I tried two or three people, and they all turned me down. This was with President Johnson's knowledge and instruction. Finally John Macy called me up and said the President would like to send a man named Tom Paine . . . I saw him and concluded that this fellow had the needed know-how in science, engineering, and management. He had the necessary judgment and capability and he was young. So I said fine, we'd like to have you. Then I asked him, "Now why do you want to do this?" He said, "I've been selected to be in the top group at GE in Headquarters, from which they pick the group vice presidents and the top people of the company, but I told them no, I want to come in the government for a few years to broaden myself." I thought that's a mighty fine recommendation.

Well, he hadn't been there but a short period of time before a fellow named [James M.] Beggs sent me word through a friend of his that if that was good enough for Paine, he'd like to do it, too. He'd been selected to be in the top group at Westinghouse from whom they picked the top people, but he'd like a tour of duty in government to broaden himself. Shortly after that, Carl Harr [of the President's staff] called me up and said I've got another one for you, Phil Whittaker of IBM. You see, the power of example is compelling. A lot of people now want to do something like Tom Paine's doing. With the change of administrations, Whittaker is going over to the Air Force to be in charge of all procurement. Beggs has become undersecretary of Transportation and Tom Paine's running NASA. Many areas of government are benefiting from the work of these able, business-trained government executives who want the broadest of experience. This is what we wanted to accomplish in moving men from the Centers to Headquarters and vice versa.

Thomas O. Paine (NASA photo.)

CHAPTER 2

THOMAS O. PAINE

(1 9 2 1 – 1 9 9 2)

Dr. Thomas O. Paine was appointed deputy administrator of NASA on January 31, 1968. Upon the retirement of James E. Webb on October 8, 1968, he was named acting administrator of NASA. He was nominated as NASA's third administrator on March 5, 1969, and confirmed by the Senate on March 20, 1969.

During his leadership, the first seven Apollo missions were flown, in which 20 astronauts orbited the Earth, 14 traveled to the Moon and four walked upon its surface. Many automated scientific and applications spacecraft were also flown in U.S. and cooperative international programs.

Paine resigned from NASA on September 15, 1970, to return to the General Electric Co. in New York City as vice president and group executive, Power Generation Group, where he remained until 1976.

Paine began his career as a research associate at Stanford University from 1947 to 1949, where he made basic studies of high-temperature alloys and liquid metals in support of naval nuclear reactor programs. He joined the General Electric (GE) Research Laboratory in Schenectady, New York, in 1949 as research associate, where he initiated research programs on magnetic and composite materials. In 1951, he transferred to the Meter and Instrument Department, Lynn,

Massachusetts, as manager of materials development, and later as laboratory manager. Under Paine's management the laboratory received the 1956 Award for Outstanding Contribution to Industrial Science from the American Association for Advancement of Science for its work in fine-particle magnet development.

From 1958 to 1962, Paine was research associate and manager of Engineering Applications at GE's Research and Development Center in Schenectady. From 1963 to 1968, he was manager of TEMPO, GE's Center for Advanced Studies in Santa Barbara, California.

Paine's professional activities have included chairmanship of the 1962 Engineering Research Foundation—Engineers Joint Council Conference on Science and Technology for Less Developed Nations; secretary and editor of the E.J.C. Committee on the Nation's Engineering Research Needs 1965-1985; member, Advisory Committee and local chairman, Joint American Physical Society—Institute of Electrical and Electronics Engineers International Conference on Magnetism and Magnetic Materials; chairman, Special Task Force for U.S. Department of Housing and Urban Development; Advisory Board, AIME "Journal of Metals"; member, Basic Science Committee of IEEE, Research Committee of the Stanford University School of Engineering, and Board of Scientific Advisors of the Quarterly Journal "Research Policy."

Paine was born in Berkeley, California, November 9, 1921, son of Commodore and Mrs. George T. Paine, USN (Ret.). He attended public schools in various cities and graduated from Brown University in 1942 with an A.B. degree in engineering. From 1946 to 1949 Paine attended Stanford University, receiving an M.S. degree in 1947 and Ph.D. in physical metallurgy. He has received honorary doctor of science degrees from Brown University, Clarkson College of Technology, Nebraska Wesleyan University, the University of New Brunswick (Canada), Oklahoma City University, and an honorary doctor of engineering degree from Worcester Polytechnic Institute.

In World War II, he served as a submarine officer in the Pacific and in the Japanese occupation. He qualified in submarines and as a Navy deep-sea diver and was awarded the Commendation Medal and Submarine Combat Insignia with stars.

In 1985, the White House chose Thomas Paine as chair of a National Commission on Space to prepare a report on the future of *space exploration. Since leaving NASA 15 years earlier, Paine had been a tireless spokesman for an expansive view of what should be done in space. The Paine Commission took most of a year to prepare its report, largely because it solicited public input in hearings throughout the United States. The Commission report,* Pioneering the Space Frontier, *was published in a lavishly illustrated, glossy format in May 1986. It espoused "a pioneering mission for 21st-century America"—"to lead the exploration and development of the space frontier, advancing science, technology, and enterprise, and building institutions and systems that make accessible vast new resources and support human settlements beyond Earth orbit, from the highlands of the Moon to the plains of Mars." The report also contained a "Declaration for Space" that included a rationale for exploring and settling the solar system and outlined a long-range space program for the United States.*

Paine was married to Barbara Helen Taunton Pearse of Perth, Western Australia. They had four children: Marguerite Ada, George Thomas, Judith Janet, and Frank Taunton. Paine died of cancer at his home in Los Angeles, California, on May 4, 1992.

Editor's note: The following are edited excerpts from two separate interviews conducted with Dr. Thomas O. Paine by Robert Sherrod. Interview #1 was conducted while Dr. Paine was administrator of NASA at NASA Headquarters on August 14, 1970. Interview #2 was conducted on October 7, 1971, in New York, after Dr. Paine had left NASA to become vice president and group executive for GE's Power Generation Group.

Interview #1

I've heard somebody say that you wanted to put a Kennedy half dollar on the Moon and it got turned down at the White House.

No, as a matter of fact that is not the kind of question that I would have raised at the White House. I did raise the question whether or not we should put a Kennedy half dollar on the Moon, and I also thought of putting one of his PT-109 tie clips, or some artifact. It seemed to me we ought to do something. And then we came around to the view that probably the best thing that we could put on the Moon as a memento of Kennedy's involvement was his statement about going to the Moon. So it got incorporated into the silicon disk,[1] and, as far as I was concerned, that was a suitable way of having a piece of President Kennedy there.

Yes. That's what he'll be remembered for, I'm sure. Do you know—I think this is true—when the language of the inscription on the lunar module (LM) [plaque] was sent over to the White House (I think a word or so was changed), Nixon knocked out his middle initial? I think that's the first time he decided to drop the M from his name.

I hadn't heard that we were responsible for the disappearance of Milhous.

Somebody over at the White House told me this.

It's possible. Actually, that plaque[2]—we had quite a flap over here in those final days before we had to make the decision and load the spacecraft about what we should carry. The decisions were prima-

Closeup showing the goodwill message disc (right) compared in size to that of a Kennedy half dollar. The small disc, which was placed on the surface of the moon by the Apollo 11 astronauts, contains statements by Presidents Eisenhower, Kennedy, Johnson, and Nixon along with messages of goodwill from leaders of 73 countries around the world. (NASA Photo S69-39148.)

rily by Shap [Willis H. Shapley, NASA associate deputy administrator] and myself. I remember being in Shap's office one afternoon—our shirts were open, our ties were loosened—and I remember getting a tablet and we sketched out this little plaque. And then we got the artist to draw it—well, we didn't like it this way, we didn't like it that way and we horsed around with it back and forth—and in a very short time we came up with the final thing. And the only change that was made in the White House was the change in tense. When we sent it over to the White House all finally done we had "We come in peace for all mankind" and they changed it to "We came in peace for all mankind." We were just delighted with that. That's what I call the kind of decision where you go in with some grandiose project to somebody and then try to get them arguing about whether it should be pink or green. The fact that they bought the whole thing was ideal, and we didn't care about changing the tense.

The other change was that they knocked out his middle initial . . .

Certainly one thing that would make the President think about how he wanted his name to go down in posterity was the fact that we were going to put it on the Moon for all time. So I'm sure it helped him make up his mind.

I remember hearing that there was a long and hard debate about whether to go ahead with the [Apollo 13] mission. Is this true?

Not with Jim Lovell. There was no debate whatsoever. But I think it's very important in discussions like this to always make sure that you've taken enough time so that you not only go through the subject once but you go through it two or three times, asking the questions in somewhat different ways, so that if he's hesitating to bring up some subject, after awhile—you continue to dig in an informal way—if there is anything there it surfaces. And I talked to Jim long enough to completely convince myself that Jim had no reservations, that he was comfortable about this flight.

As you know, I've made it a practice to go down to the Cape and talk to the fellows before every mission in an informal way, just making sure that they don't have any possible reservation that they've hesitated to bring up with Deke [Slayton] or Chris [Kraft] or Bob Gilruth, Rocco [Petrone], Sam Phillips (in his day)— just to make sure that they know they have this channel. Nothing has ever been turned up in these meetings. I always made one other point with them, and that was to urge them to feel free at any time, if they didn't feel comfortable with the way the mission was progressing even if they couldn't put their finger on all the specifics, to feel very comfortable about aborting the mission. And I've assured them that in the event that they made a decision to bring the ship back prematurely, that if they then requested the opportunity to fly on another mission immediately afterwards—I won't say the next flight, but immediately afterwards—that I would see to it that we bent all the rules and gave them this opportunity. I wanted them to feel that if they didn't feel they were on a good mission that—but they still wanted to fly a mission—that in their

Approved by Administrator Paine, here is the Apollo 11 commemorative plaque showing its attached position on the ladder of the landing gear strut on the lunar module descent stage between the third and fourth ladder rungs from the bottom. (NASA Photo S69-38749.)

own minds they would be ready and willing to turn back and we'd give them another shift. They could fly that out and they wouldn't have to go to the end of the line, they wouldn't lose their chance to go to the Moon.

Are you speaking now of before the launch or during Earth orbit or . . .

Before the launch. This is an informal thing, and has traditionally been my last statement to the astronauts. For example, I would usually go down and have dinner with them. And we talked at

21

dinner usually about other subjects, and finally got around to the mission over dessert and coffee. We usually start out with a few drinks beforehand to get everybody relaxed. And then my last statement, in private, to the three astronauts together, will be to say "I have one final statement that I want you to remember as you fly this mission," and then I would make that offer to them, that they should feel free to bring the ship back if they don't feel that they want to continue with the mission, and we'll give them another crack at it.

Did you start this with Apollo 8?

Yes.

How about Apollo 7?

I don't remember Apollo 7. I don't believe I did with Apollo 7. I believe I started with Apollo 8. Apollo 7, of course, was a somewhat different mission because once the thing was launched, you know, they could bring it back anytime, and we wouldn't want to repeat that mission anyway. The whole objective was to go there and check the spacecraft—it was a little different. But Apollo 8 and sub-sequently, this offer was made. Also, before I made the decision to send Apollo 8 around the Moon I called Frank Borman on the tele-phone and discussed it with him, as a part of the final decision process, to make sure I understood how Frank felt about it.

. . .Do you ever recall using the LM as a lifeboat—being apprised of that use of the LM as a lifeboat—during the whole Apollo 8 discussion?

The fact of the matter is that the whole reason for sending Apollo 8 around the Moon was because the LM wasn't ready. And this was the thing which came out in August while George Mueller and Jim Webb were overseas. The question that came up was that all of a sudden—and this has always seemed very mysterious to George— the LM schedule suddenly slipped a couple of months the minute he left the country . . . As soon as they got over the Atlantic, the LM schedule was re-examined and found to be a couple of months late. Then the question came up what should we do? Should we postpone the next Saturn V launch and move it on out until the LM was available and go ahead and fly the mission (which we subsequently flew as Apollo 9, the Earth orbital LM mission). Now, from the standpoint of nicety, I think you could make a very good case that, yeah, the first thing to do is fly the LM in Earth orbit before you send it out to the Moon, and then send the two out to the Moon as we did in Apollo 10. But it wasn't—there was no question from the beginning of that decision about having a LM along, and the whole reason we were going out to the Moon is because we didn't have a LM. And the discussion was—we did raise the discussion—to what degree does this increase the risk of going to the Moon? And one of the things we had to look at was the fact that, well, if we had a LM along as we would in later missions, we would have a return capability. This was partly offset by the fact that we were going to fly a free-return trajectory, so it was only when we dropped into lunar orbit that we would lose this—we would be back to the single-point failure. On the other hand, once we dropped into the lunar orbit on a regular Apollo flight we also lost the LM, you see. So in looking at it, the only time that the LM would have helped you was in the time that we had the failure in Apollo 13 on the long coast out to the Moon. And this was a risk that we just plain tried to assess and decide whether or not to fly; we decided to accept that risk.

I can't find anywhere that anybody ever said, my God, if we fly this Apollo 8 without a LM we are going to lose the lifeboat capability.

I do recall the subject coming up but I can't remember when, it may have been as early as August [1968] or in some later discussion. But it wasn't given much, we certainly didn't put much on it. And the reason was that we only envisioned the lifeboat as being a viable thing in that relatively brief period on the trans-lunar injection and subsequent coast.

Now I think the other comment you have to make is what were the things that really concerned us. Things that really bothered us were things like the fuel cells. We were very concerned about what would happen if we lost power at various stages en route. And the other thing that bothered us was the life support system. What would happen if we lost oxygen and so forth? I think the other point is that during the Apollo 9 mission we did fly the lifeboat mode to check out the LM and make sure that with the two spacecraft mated that there would be no unforeseen vibration, torsion, or other problems that would occur while using the descent engine in that configuration. The one configuration we'd never flown before was the configuration that we had in Apollo 13 just before reentry, where we undocked the LM and then we undocked the service module and for awhile, we had the LM attached to the command module. That was a mode we had not flown before and there was some concern.

Have you thought about the *Life* contract in its origins? If you had it to do over again . . . do you think you'd do it as it was done then?

I certainly don't think I would do it again. I think that it was not a good move to make, more on principle than on anything else. Now I did look back recently and asked the question and worked with Julian Scheer [NASA's public information officer] to see if we could, in a dispassionate way, ask ourselves has this hurt the agency. And I think if you look at it dispassionately, you have to concede that it did not hurt the agency. Now, God knows there was a tremendous amount of complaint . . . It's been very annoying for us. And an awful lot of the other media people have been most upset and have complained about it. I think looking back on it, what they really sold was their byline in the family stories, and as far as NASA was concerned, what happened was that we got top billing in *Life* magazine on this. And I think *Life* is so much the standard for this sort of thing . . . it was really the leading publication, and I don't think we could have gotten a better shake. And when I look at what all the other people did to cover the program, you can't show that any daily or any weekly didn't give us absolutely full coverage. They couldn't have a byline from the astronauts but look at what all the picture magazines have done

24

with our pictures. I think they're terrific. I think you have to honestly say that it hasn't hurt the agency. But I must also tell you that I would never have signed such a contract, on the principle that I don't think government employees, who are paid by the public, should have done this . . .

Interview #2

Among the procurements made for Apollo, looking back, were there any times you thought a mistake was made . . . when you thought to override the system?

. . . The only one that I think I ever called wrong was the one that was the most logical one we ever called. It was Boeing and General Motors—who came in with the lunar rover contract against Bendix. By God, we went through that competition and everything said that it should go to Boeing and General Motors. They had done everything—lower price, everything pointed to it—and I knew damn well it wasn't right. I knew damn well Bendix was a better outfit and that it would stay within the price and would give us a better product . . . We had all kinds of cost overruns. We finally had to take it away from the Boeing crew that had it and have Boeing send in a whole new crew. All the things that we knew were going to happen at the management level did happen and yet it was too big a move to override it. Everything pointed to giving it to the people that we gave it to. It would have had to be almost an illogical act [to give the contract to Bendix] and yet the system produced the wrong answer . . .

I bet you were very glad to see the lunar rover work as well as it did.

I sure was. I was scared to death of that contract. As I said, I think that was the worst contract that I was ever in on. That was the one decision that I would have reversed if I had a bit of hindsight. I was awfully glad that thing worked. I was scared to death it wasn't going to work. If that thing hadn't worked, with the TV camera on it and in front of the whole world—if it had just sat there [on the Moon] after all we'd done to take it there—it would have been a fiasco, a terrible fiasco.

Interview #1

What were the circumstances behind your resignation?

I had a fellow come in from another company—Xerox, as a matter of fact—and say that they had a high-level opening and that they wanted to know would I be interested in leaving Washington. It was a pretty good job, although I wasn't really interested in it. While—I always make it a practice never to turn anybody down, [but tell them I'll] think about it for a few days—and while I was thinking about it I ran into one of my old friends from GE and he asked me was there any possibility of my leaving, and I said no, that I was pretty well set, except that somebody had made me an offer. Sort of like a girl with a new hat, you know—it makes you feel good inside, makes you pleased even if it isn't a good one—you are reassured that you are still respected. And he said, gee, as a matter of fact they were thinking of some fairly major reorganization at GE and before I took any other offer they certainly hoped I would re-contact them. And I made my classic statement that, I really didn't think I would ever go back to GE; I'd probably want to do an independent company presidency and I really didn't think they would have anything that would interest me. Apparently it made an impression on them, because they came back shortly thereafter and said, "Look, we really don't want you to go anywhere else, we really want you to come back, we're short of top management people, and in your age bracket and experience there almost isn't anybody, and you would certainly be in a very good position, and we think we could give you a job that would give you some national problems that would have public interest associated with them and at the same time pay you a good salary." So I said, "I don't think I'd be interested but make me an offer." And as I thought about it, more and more it seemed to me—as I said to the president—that maybe this was a good time for a transition here at NASA. I mulled it over for a couple of weeks, and then that Saturday before the announcement decided that I would indeed go ahead with it. One of the things I've learned from Julian Scheer is that the only way to keep it a secret is to tell everybody, and then you never get your secrets divulged because you essentially don't have any. And it seemed to me that I should do this. I talked to the president of GE about this, Fred Borch, and he agreed that although at his end he still hadn't talked to the people who would be affected and who would be essentially replaced when I showed up, that he agreed that it would be a damn good thing for me not to be in a position of having accepted an offer from him and then keeping it to myself. He said "Why don't you just get it out right away but don't say what you're going to do here?" So on that basis we did it, and I went ahead on Saturday—as soon as I made the decision—I went ahead and called the White House (the President was out in San Clemente) and said I wanted to see him Monday or Tuesday. I then called a small group of people here in the office Sunday and said, "We've got a lot of work to do: we've got to announce my resignation, I want to do it properly and I want to send out a good letter to people in NASA telling them why I'm doing it; I want to notify the people in Congress and I want to notify top executives here . . . ."

George Low didn't know?

No, and I debated that. I talked to George [Low] Saturday, right after I made the decision, and George was just leaving and he was full of his trip. And I decided, gee, George isn't going to do anything about this anyway. If I tell him, instead of enjoying his trip up from Houston he's going to be thinking about nothing but this all the way, and I thought why don't I wait and I'll let him know just before I see the president? Normally I wouldn't do that, but this was such a special circumstance, and there wasn't anything he could do about it. So I did tell him, "George, I want you to call in at noontime every day, then in the evening when you find a place

to stay call the NASA operators and tell them where you're staying," I said, "because there is something I'm going to want to tell
you while you're en route." And George afterwards told me,
"When you said that, Tom, my guess was that you were going to
announce your resignation." [Laughs.] So I thought that was pretty
good of George.

He had some warning after all.

Yes.

**How much do you think your two-and-a-half years of government
service have cost you?**

Well, it's a hard question to answer because you have to ask what
I would have done and how that would have come out. My guess
is it hasn't cost me anything . . . I think you have to ask yourself
the question "what really could you have done better with your
life?" And the answer to that has to be "absolutely nothing." This
has been a wonderful opportunity to work with a bunch of great
people. I remember when I first got the job as a deputy—way back,
even before I reported to Washington—Bill Pickering called me
on the phone and said, "Gee, you know I'd like to get to know you,
and once you get East you're going to be pretty busy. Why don't
you come down and spend a day at the lab before you go back
there?" And I said, "Gee, I'd like to do that." He said "I tell you
what: we'll take you out to Goldstone and let you look around."
And Barbara [Paine's wife] said she wanted to go too, and he said
fine. So Bill sent a little light plane up and we flew out to
Goldstone. On the way we stopped at Edwards and we stooged
around the hangars and talked to the fellows there. They were a
fascinating bunch of people, and the old B-70 was coming in and
making some touch-and-go landings (that's a pretty spectacular
thing to watch) and everybody was wearing informal clothes and
really fascinated by what they were doing. And we went out to
Goldstone . . . We got back and Barbara, I remember, was
absolutely ecstatic, and she said "I've been there before: these are

all the young people we worked with during the war," she said. "These are the RAF types who have just come back from battling the Luftwaffe, these are the young submariners in Perth, this bunch of real selfless people that are doing what they want to do and taking risks and fighting odds and really doing exciting things." It was absolutely true. I think for me this has been a great recharging of my battery, and people talk about the awesome responsibility. God knows it's there—it gnaws at you all the time, there are nights when you don't sleep. I think there are very few jobs in the country—certainly none that I know of outside the military, and really not many there even—where a guy has to so put his neck on the guillotine, in the full view of the entire world, for the prestige of the country and the fates of some good people, and the professional qualifications of hundreds of thousands of people all laid on the line.

I've heard that you were vice chairman of the Scientists Against Goldwater in 1964.

That's not quite right. I had never been particularly active politically any place I'd lived, and my registration had normally been whatever I thought was the best local party and whoever I thought really in my vote might count for the most in the primary . . . When we lived in Massachusetts I actually registered Republican one year when Ike was running against Taft for the nomination, because I was so anxious to have Ike nominated instead of Taft and move the Republican Party over to a more liberal stance. Normally, I've always registered as a Democrat, and have considered myself a Democrat except for that one registration as a Republican to vote against Taft.

When the campaign came out in '64, where Goldwater was running against Johnson—like many other people who were associated with science—I became very concerned about the casual way that Goldwater was talking about the use of nuclear weapons. Really that was the one thing (I didn't really approve of his conservative position, but I wouldn't have bothered to oppose him except by my vote—except for this rather emotional issue which

29

got raised with respect to nuclear weapons). I organized, and was the chairman of the Santa Barbara Chapter of the Scientists and Engineers for Johnson. It was a one-man effort. We rented an office and we got a dozen people or so to go around and write letters asking for contributions. We took the money that was contributed and ran ads in the newspaper in Santa Barbara saying it was important not to drop nuclear weapons indiscriminately. It was a pretty casual thing . . .

He's [Goldwater] a great guy. I like him and I have great respect for him. He's a man of principle, and I think I was perhaps a little swept away, but that is such a basic issue—I just didn't think as a citizen who knew about the power of nuclear weapons I could sit on the sidelines while this sort of thing went on. But personally I like Barry—he's a great guy and he's been very good for the space program.

Did Johnson ever recall after you came to NASA that you had been chairman of the Santa Barbara . . .

No, he never knew anything about it. It was strictly a local Santa Barbara thing.

Interview #2

One thing that struck me was that, unlike Jim Webb, who had his Senator Mondale and his Congressman Ryan and his George Mueller and many others, I can't find that you really made any enemies during your time there [as NASA administrator].

I didn't make many enemies, I don't believe, but this was partly due to the time. I think given a little more time I certainly could have and would have, but NASA was going through its very greatest testing and greatest trial, and at a time like this everybody was getting aboard the bandwagon.

Interview #1

How did you decide on George Low for deputy [administrator of NASA]?

Basically it involved canvassing the whole country. I looked in universities, I looked in industry and I looked within NASA for candidates, and we compiled a fairly extensive list of candidates. And frankly, I must also state that one of the things in my mind was the fact that I was a bird of passage in the Nixon administration and I would be leaving one of these days—although I didn't realize how soon it would be—and I was very anxious to get somebody in who was fully qualified to run the agency in my absence, and also somebody who would give the president wide latitude in his choice of a successor. It seemed to me that while we were trying to finish up the Apollo Moon landings that having a person like myself in with a strong technical background and a person who'd exercised technical judgement over many years in a wide variety of areas . . . that this was a damn good thing to have to make that Apollo Program come out right . . . It seemed to me that whoever we put in as Deputy should leave the president the option of putting another Jim Webb kind of guy in as administrator, which I think is a good combination to run the agency.

The other criterion I put on it was I was very anxious to get a younger man. I had a feeling that when you give a person in his 40s an opportunity to run something like NASA you really bring out the best in NASA and you bring out the best in the guy. If you go and get a retired fellow who is in his 60s or late 50s, and ask him to do it, well, he's already got his reputation made and all he can do is lose and so he's going to be a little too cautious. He's not going to work the hours, he's not going to have the drive, the adventuresomeness, it won't be such a big thing to him. I was most anxious to get somebody preferably in his younger 40s but certainly no older than his late 40s. And that cut out a lot of very talented people . . . I looked at people, for example, like Frank Borman, who I felt would have in many ways been a good deputy. I looked at people like Sam Phillips. I felt it would be a damn good idea not to get a military man. It seemed to me that although we had to get along very closely with the Pentagon, the president ought to keep the separation between NASA and the Pentagon—we ought to be the peaceful guys and they ought to be the military guys—and not mix them up. We shouldn't be a paramilitary organization. As far as Frank went—I thought Frank would be wonderful as maybe even the administrator, the Mr. Outside, the Jim Webb if you like—but it seemed to me that as deputy you really wanted a guy who could run the thing, although I certainly toyed with the idea of Frank and felt that in many ways he would have been a darn good choice, particularly if I were to stay on as administrator indefinitely. I felt that would be a good combination, but that probably in the long run it ought to be the other way around—the deputy ought to be Mr. Inside—a guy with a high technical competence— and the Administrator ought to be Mr. Outside, the fellow who goes to the Washington dinners and the White House and so forth, and that really you could make a better case almost for Frank in that role than you could as deputy. And I talked to various people about their interest, and I also looked very carefully among the top scientists of the nation and actually approached a couple of people like Charlie Townes[3] and asked them whether they would be willing to come in, because I felt that the other way you could run the agency would be to have a fellow like myself, who of course you could replace with another guy like me in the administrator job, and get a real top scientist in, a Nobel Prize winner, as the deputy, and assure yourself then that the program really is oriented to science to a great degree.

What did Townes say?

Charlie is having too fine a time out at Berkeley and wasn't about to make the move. And I can understand that. Charlie, again, was not ideal, though. He was a little older than I'd like to have seen, he'd kind of made his reputation. It had a little bit the gloss of a circus thing. On balance, after thinking it over, talking to a number of people, I finally reached the judgment that George Low really was the ideal choice.

Wernher von Braun talks with NASA Associate Administrator Dr. Thomas Paine at the Launch Control Center of the Kennedy Space Center prior to the launch of Apollo 6 on April 4, 1968. (NASA Photo 107-KSC-68P-125.)

Did you consider Frank Borman long enough to talk it over with him?

I certainly considered him long enough to talk it over with him, but I did not talk it over with him as I recall.

I just wondered what he said. You waited a long time. You waited six months to pick a deputy.

Yes, I did. That partly reflects the fact that I wanted to get a very good candidate because I knew when I went over to the White House this appointment, which is a presidential appointment, would have to be looked at very carefully. And I didn't want any premature moves there. I wanted to have somebody that I'd fight and bleed and die for. I don't believe in losing battles.

I heard there were some people in the White House who said, "Well, who is George Low? Whoever heard of George Low?"

Well, I certainly had to sell him to Peter Flanigan [an assistant to President Nixon] and to Lee DuBridge [President's Nixon's science advisor]. Lee, of course, would like to have seen a prominent scientist. Peter Flanigan, I think, wanted to assure himself that this was indeed going to be a good appointment by the president.

Interview #2

. . .It seems to me that during your career at NASA—you can correct me on any of this—the highlights of your administration, insofar as you were personally concerned or as far as the outside world, were the C Prime Decision [Apollo 8], Apollo 13, and then the fact itself of the lunar landing with the Abernathy incident.[4] What else would you put in there?

Well that's a hard question. There are an awful lot of things. Of course there are a lot of things that you don't really see with much prominence that nevertheless are fairly important. I think that at the end of my regime there were some things that had to be done that I would put fairly high on my list. First of all, there was the business of getting through the transition to the Nixon administration and somehow capturing the Nixon administration for space. Nixon had made some very negative statements about space, and I think he had always associated it with his arch-rival Jack Kennedy. Getting us through that transition and getting a fairly solid base in the new administration, I felt, was one of the very great challenges . . . Having Frank Borman closely associated with the president was quite a help during this period. Bringing George Low aboard gave me the ability, then, to turn NASA over to whoever would be the next administrator knowing that we had some good solid technical competence at the top level, and that no matter who the president put in alongside, even a politician with very little background, that NASA would still have the right kind of advice at the top.

Working within the new administration to get the new science adviser, Lee DuBridge, and the new secretary of the Air Force, Dr. Seamans, all pulling together to give us the role we needed through the Space Task Group and then defining for the president a good solid Nixon space program and get him to come out and publicly espouse it while at the same time sell it to the Democrats who were running the Congress—these were all great challenges.

. . . Then we had to preserve the Skylab program when it was being threatened by the Air Force's MOL [Manned Orbiting Laboratory] program.[5] To thread our way through that whole period was a very tricky thing. It took an awful lot of thought, patience, and doing, along with moments of wondering whether or not we were going to pull it off.

Who talked Nixon into killing MOL . . . did you?

No. I had to, of course, play that very straight. I couldn't be a double agent who, on the one hand, would publicly advocate continuing the program but privately would try to kill it. At the same time, I was absolutely determined that if one of those programs survived—and I was damn sure that only one would survive—I wanted it to be the Skylab. I was very anxious that this be done for a number of reasons. I thought the nation would get more out of it, and I looked forward to a Soviet/U.S. cooperation in this era if we went ahead with Skylab... I thought [it] would turn out to be exactly the opposite—a rather pernicious competition—if it went the other way. It seemed to me that an awful lot was riding on it. Furthermore, it was a program that I thought would have enormous interest in the post-Apollo area. Once we'd been to the Moon, then it seemed to me probably the first thing we'd occupy was orbital space, and that we needed a sort of transition: let us have a program there that would attract the public's interest . . . There were a lot of considerations and a lot of things to work around in that period. Not least of which, of course, was keeping the U.S. manned space effort going.

. . . There were other problems in between, such as the care and feeding of astronauts, the ability to handle a very delicate public relations program with Julian having to say no to a lot of people and make a lot of enemies, which is a very useful thing for an administrator to have somebody like that who is willing to take the negative side of things and keep the administrator from having to be the heavy character . . . I well remember when I first came aboard and went down for my first launch at the Cape—the first big launch, which was the Apollo 6, the last of the unmanned launches in the Apollo Program. The air of animosity and the frank feeling I got from the reporters that the best story they could possibly get would be to have the damn thing explode on the pad the next day, in which case they would have credit lines on the front page, whereas if the damn thing went up, well there wasn't much of a story in that. And they also knew that we didn't know what the hell we were doing and that we were a bunch of bums and that it was all a great facade really—the relations with the press were not good when I first started out.

ENDNOTES

1. A small disc carrying statements by Presidents Eisenhower, Kennedy, Johnson, and Nixon and messages of goodwill from leaders of 73 countries around the world was carried aloft by the Apollo 11 crew and left on the surface of the Moon. The disc also carried a listing of the leadership of the Congress in 1969 and a listing of members of the four committees of the House and Senate responsible for NASA legislation. In addition, the names of NASA's top management, including past administrators and deputy administrators along with the present NASA management [as of 1969] was included. The disc, about the size of a 50-cent piece, is made of silicon. Through a process used to make microminiature electronic circuits, the statements, messages, and names were etched on the gray-colored disc. The messages were first photographed and the photo reduced 200 times to a size much smaller than the head of a pin (0.0425 x 0.055 inches). The resulting image was transferred to glass which was used as a mask through which ultra-violet light was beamed onto a photo-sensitive film on the silicon disc. After a photodevelopment step, the disc was washed with hydroflouric acid which accomplished the final etching which appears on the disc as a barely visible dot. NASA's Electronics Research Center at Cambridge, MA, was assisted by Sprague Electric Company's Semi-Conductor Division, Worcester, MA, in preparing the disc. At the top of the disc is the inscription: "Goodwill messages from around the world brought to the Moon by the astronauts of Apollo 11." Around the

rim is the statement: "From Planet Earth – July 1969." The messages from foreign leaders congratulate the United States and its astronauts and also express hope for peace to all nations of the world. Some are handwritten, others typed, and many are in native language. A highly decorative message from the Vatican was signed by Pope Paul. Silicon was chosen to bear the miniaturized messages because of its ability to withstand the extreme temperatures of the Moon which range from 250 degrees to minus 280 Fahrenheit. The disc itself is fragile and was transported to the Moon by the astronauts in an aluminum capsule where it remains on the lunar surface today. The words on the disc, although not visible to the naked eye, remain readable through a microscope. "Apollo 11 Goodwill Messages," NASA News Release No. 69-83F, July 13, 1969, pp. 1-4.

2. Beginning with Apollo 11, each crew carried to the lunar surface a small commemorative plaque attached to the ladder of the landing gear strut on the lunar module descent stage between the third and fourth ladder rungs from the bottom. The Apollo 11 plaque was signed by President Nixon and the three Apollo 11 astronauts. It bears the images of the two hemispheres of the Earth and the following inscription:

HERE MEN FROM THE PLANET EARTH
FIRST SET FOOT UPON THE MOON
JULY 1969, A.D.
WE CAME IN PEACE FOR ALL MANKIND

The plaque is made of #304 stainless steel measuring nine by seven and five-eighths inches and one-sixteenth inch thick. It weighs approximately one pound and thirteen and seven-eighths ounces. The finish has the appearance of brushed chrome and the world map, message, and signatures are in black epoxy which fills the etched inscriptions. To fit properly around but not touching the landing gear strut and to allow room for the insulation material which covers much of the lunar module, the plaque was bent around a four-inch radius. Covering the plaque during flight is a thin sheet of stainless steel which was removed by Armstrong during EVA activities on the Moon. The plaque was made at NASA's Manned Spacecraft Center. "Apollo 11 Flags," NASA News Release No. 69-83E, July 3, 1969, pp. 1-4.

An alternate account of the origins of the commemorative plaque is given by Jack A. Kinzler, a model maker and engineer who worked at the Johnson Space Center during this period. In an interview conducted with Kinzler on April 27, 1999, by the Johnson Space Center Oral History Project, Kinzler describes how he developed the idea of the plaque, produced the prototypes, and delivered the finished product at the Johnson Space Center.

In response to a high level meeting at JSC, Center Director Bob Gilruth phoned Kinzler asking him for ideas on how best to celebrate the first lunar landing. Kinzler responded with the idea of a stainless steel plaque that could be attached to the LM. Prototypes were made by both Kinzler and co-worker David L. McCraw and carried over to the meeting for approval. Everyone liked the idea and Kinzler was given the go ahead to refine the prototype into a finished product.

According to Kinzler, "the first conceptual metal plaque that I took into the meeting had the American flag engraved in the middle and it was painted with red, white, and blue paints that were baked into the etched definition. It then had a place for a message, the name of the landing spot, and crew signatures." Gilruth looked at the plaque and suggested to Kinzler that they remove the flag and replace it with the two hemispheres of the Earth, one representing the Eastern Hemisphere and the other representing the Western. Kinzler went on to explain that Gilruth said "If we do that, any creatures that come in future years and look at the lunar module that's sitting there on the surface will recognize the source from where this device came."

"Once the plaque concept was approved," Kinzler said, "NASA Headquarters took it over and said 'We'd better go by the President and see if he approves this sort of thing.'" President Nixon liked the idea and wanted his signature to be included on the plaque. "NASA called down to us," and said, "we have this picture you're going to use to create your final metal plaque. Can you add the President's signatures to it?" Kinzler replied, "Yes, but we need a faxed copy of his signature." So Headquarters sent a copy of his signature to JSC. Jack Kinzler's secretary, a Mrs. Germany, saw the faxed signature that was sent to his office and pointed out that it did not look correct. According to Kinzler, Mrs. Germany's mother-in-law received a letter at that time congratulating her on her reaching her 100th birthday. The letter was signed by President Nixon but his signature did not include his middle initial "M" which was included in the faxed signature. After carefully comparing the two signatures, Kinzler decided to call NASA Headquarters to try and confirm which one they should use on the plaque. Headquarters agreed that he should use the more recent signature and, as a result, the one used on the plaque is that copied from Mrs. Germany's mother-in-law's letter received from the President. (Source: Jack A. Kinzler interview conducted for the NASA Johnson Space Center Oral History Project at JSC on April 27, 1999 by Roy Neal.)

Julian Scheer, assistant administrator of public affairs for NASA from 1962-1971, offers additional details concerning the wording of the lunar plaque. According to Scheer, President Nixon insisted on making a change to the original plaque wording which read "We Came in Peace for All Mankind." The President wanted to insert "Under God" after the word "Peace." However, because the timing of the change had come so late after the plaque had been installed on the spacecraft, the requested change could not be made and the original wording of the plaque, to this day, remains on the lunar surface. (Source: "What President Nixon Didn't Know" by Julian Scheer, an online article that appeared on July 16, 1999, as a special to *space.com* at URL *http://www.space.com/news/a11plaque.html*)

3. Charles H. Townes was trained in physics at Duke University and specialized in the development of laser and maser technology. He first worked for the Bell Telephone Laboratories and, in 1948, joined the faculty of Columbia University, leaving there in 1961 to move to the Massachusetts Institute of Technology and on to the University of California. For his work on the maser, Townes received the Nobel Prize in 1964. See David E. Newton, "Charles H. Townes," in Emily J. McMurray, ed., Notable Twentieth-Century Scientists (New York: Gale Research Inc., 1995), pp. 2042-44.

4. Prior to the launch of Apollo 11, a large group of the "Poor People's Campaign," accompanied by muledrawn wagon, approached the Kennedy Space Center led by the Reverend Ralph Abernathy and Hosea Williams of the Southern Christian Leadership Conference. The marchers stopped at the visitors center where the reverend intended to protest the launch, feeling that the money used to finance space exploration should go to the poor. Paine and Julian Scheer, NASA's public information officer, met the demonstrators. Seeking to extinguish a potential PR nightmare, Paine explained to Abernathy's group that he was a member of the National Association for the Advancement of Colored People and sympathetic with the poor, but stopping the launch would not put money in their hands. Paine offered to admit a delegation to observe the launch. NASA stationed buses in Cocoa, 15 miles away, to pick up the delegation and placed food packets on the seats. When Apollo 11 launched the next morning, 100 men, women, and children of the "Poor People's Campaign" were among the guests watching the spectacle. Even Abernathy was briefly overcome by emotion when interviewed on television. Paine's timely intervention helped relieve a delicate situation. Gordon L. Harris, *Selling Uncle Sam* (New York: Exposition Press,1976), pp. 110-111.

5. MOL (Manned Orbiting Laboratory) was a USAF manned space program that originated in the early 1960s in response to maintaining the ultimate "high ground" for military interest. As envisioned, the Earth orbiting station would accommodate a crew of two along with an attached modified Gemini spacecraft and retrofire package for ascent and reentry. The Gemini/MOL stack would be launched manned into orbit aloft a Titan III-M booster where crews would work in a shirt-sleeve environment of either nitrogen and oxygen or helium and oxygen. Designers planned to use fuel cells developed for Gemini and Apollo that would allow crews to work a 30-day mission, while longer stays would have been powered either by solar panels or thermoelectric radioisotope generators. Planned experiments performed aboard MOL ranged the gamut from military reconnaissance using large optical cameras and side-looking radar through interception and inspection of satellites, to exploring the usefulness of man in space and tests of Manned Maneuvering Units. At the end of each MOL mission, crews would climb aboard the attached Gemini through the use of a unique rear entry hatch. The retrofire package would then send the crew back to splashdown on Earth. On June 10, 1969, the MOL program was officially cancelled. Many of the reconnaissance systems ended up in later KH series satellites, while some of the manned experiments were accomplished on Skylab. Donald Pealer, "Manned Orbiting Laboratory—Part 1," Quest Vol. 4 No. 2 1995, pp. 4-16.

Wernher von Braun, the Marshall Space Flight Center's first director (1960-1970), in his office in Huntsville, Alabama, on September 16, 1960. (NASA Photo, available on NASA Image Exchange: *http://www.nix.nasa.gov.*)

CHAPTER 3

WERNHER VON BRAUN

(1 9 1 2 – 1 9 7 7)

Wernher von Braun was one of the most important rocket developers and champions of space exploration during the period between the 1930s and the 1970s. The son of a minor German noble, Magnus Maximilian von Braun, the young space flight enthusiast was born in Wilintz, Germany, on March 23, 1912. As a youth he became enamored with the possibilities of space exploration by reading the science fiction of Jules Verne and H.G. Wells, and from the science fact writings of Hermann Oberth, whose 1923 classic study, Die Rakete zu den Planetenräumen *(By Rocket to Space), prompted young von Braun to master calculus and trigonometry so he could understand the physics of rocketry.*

From his teenage years, von Braun held a keen interest in space flight, becoming involved in the German rocket society, Verein fur Raumschiffarht (VfR), as early as 1929. As a means of furthering his desire to build large and capable rockets, in 1932 he went to work for the German army to develop ballistic missiles. When Hitler came to power in 1933, von Braun remained in Germany, joined the Nazi party in 1937, and became a member of the SS three years later, all the while continuing to work for the German army.

While engaged in this work, on July 27, 1934, von Braun received a Ph.D. in physics. Throughout the 1930s von Braun continued to develop rockets for the German army, and by 1941 designs had been developed for the ballistic missile that eventually became the V-2. The brainchild of Wernher von Braun's rocket team operating at a secret laboratory at Peenemünde on the Baltic coast, this rocket was the immediate antecedent of those used in space exploration programs in the United States and the Soviet Union. A liquid propellant missile extending some 46 feet in length and weighing 27,000 pounds, the V-2 flew at speeds in excess of 3,500 miles per hour and delivered a 2,200 pound warhead to targets at a maximum distance of 200 miles away. First flown in October 1942, it was employed against targets in Europe beginning in September 1944. On October 6, for instance, more than 6,000 Germans deployed to Holland and northern Germany to bomb Belgium, France, and London with those newly developed two V-2s.

Beginning on September 8, 1944, these forces began launching V-2s against allied cities, especially Antwerp, Belgium, and London, England. By the end of the war, 1,155 had been fired against England and another 1,675 had been launched against Antwerp and other continental targets. The guidance system for these missiles was imperfect and many did not reach their targets, but those that did struck without warning.

By the beginning of 1945, it was obvious to von Braun that Germany would not achieve victory against the Allies and he began planning for the postwar era. Before the Allied capture of the V-2 rocket complex, von Braun, along with 500 of his colleagues, surrendered to the Americans bringing along with him numerous plans and test vehicles. For 15 years after World War II, von Braun would work with the United States army in the development of ballistic missiles.

Because of the intriguing nature of the V-2 technology, von Braun and his chief assistants achieved near celebrity status inside the American military establishment. As part of a military operation called Project Paperclip, he and his "rocket team" were scooped up from defeated Germany and sent to America where they were installed at Fort Bliss, Texas. There they worked on rockets for the United States army, launching them at White Sands Proving Ground, New Mexico, under the direction of General Electric Hermes personnel. In 1950, von Braun's team moved to the Redstone Arsenal near Huntsville, Alabama, where they built the Army's Jupiter ballistic missile, and before that the Redstone, used by NASA to launch the first Mercury *capsules.* In 1960, his rocket development center transferred from the army to the newly established NASA and received a mandate to build the giant Saturn rockets, the largest of this family of launchers eventually put an American on the Moon. Von Braun became director of NASA's Marshall Space Flight Center.

Von Braun also became one of the most prominent spokesmen of space exploration in the United States in the 1950s. In 1952, he gained note as a participant in a major symposium dedicated to the subject and burst upon the nation's stage in the fall of 1952 with a series of articles in Collier's, a major weekly periodical of the era. He also became a household name with his appearance on three Disney television shows dedicated to space exploration in the mid-1950s.

In 1970, NASA leadership asked von Braun to move to Washington, D.C., to head up the strategic planning effort for the agency. He left his home in Huntsville, Alabama, but in less than two years he decided to retire from NASA and to go to work for Fairchild Industries of Germantown, Maryland. He died in Alexandria, Virginia, on June 16, 1977.

Editor's note: The following are edited excerpts from two interviews conducted with Dr. Wernher von Braun. Interview #1 was conducted on August 25, 1970, by Robert Sherrod while Dr. von Braun was deputy associate administrator for planning at NASA Headquarters. Interview #2 was conducted on November 17, 1971, by Roger Bilstein and John Beltz.

Interview #1

In the *Apollo Spacecraft Chronology*, you are quoted as saying "It is true that for a long time we were not in favor of lunar orbit rendezvous. We favored Earth orbit rendezvous."

Well, actually even that is not quite correct, because at the outset we just didn't know which route [for Apollo to travel to the Moon] was the most promising. We made an agreement with Houston that we at Marshall would concentrate on the study of Earth orbit rendezvous, but that did not mean we wanted to sell it as our preferred scheme. We weren't ready to vote for it yet; our study was meant to merely identify the problems involved. The agreement also said that Houston would concentrate on studying the lunar rendezvous mode. Only after both groups had done their homework would we compare notes. This agreement was based on common sense. You don't start selling your scheme until you are convinced that it is superior. At the outset, neither Houston nor Marshall knew what was the best approach. And the fact that Houston happened to study the lunar orbit rendezvous mode was purely coincidental. That mode didn't even originate in Houston—it was first proposed by John Houbolt of the Langley Center. The problem with Houbolt's original study was that his weight figures for the lunar module were based on certain operational assumptions that Houston considered absolutely inadequate for the mission. So, as Houston added realistic requirements to the Houbolt scheme it lost a lot of its original charm. In the end, everybody wondered whether the lunar orbit rendezvous mode would still look attractive by the time the necessary realism had been instilled in it.

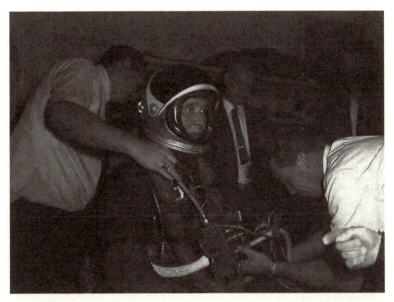

Wernher von Braun in space suit and diving equipment in the neutral buoyancy simulator at Marshall, November 14, 1967. (NASA Photo, available on NASA Image Exchange: *http://www.nix.nasa.gov.*)

At that time, as I recall, you said the way it was originally presented by Houbolt, that the lunar module was not even pressurized. The two astronauts would simply fly down to the lunar surface in their space suits.

That is correct. The first lunar orbit rendezvous scheme that I was exposed to envisioned a LM without a pressurized ascent stage. The two astronauts were to ride down in their space suits and there was to be no cabin pressurization. This was the first item that Houston said was unacceptable, and so pressurization was added. As a result, the weight of the LM went up. The next thing was that the guidance system as originally identified was considered inadequate, both for the descent and the ascent. Somebody overstated the situation a bit by saying the original system consisted of a "recticle and a stopwatch." A very sophisticated guidance system was put in the LM to replace it. Then, another redundant guidance system was added. All of this, of course, further increased the weight of the LM. The crux of the problem was this: John Houbolt argued that if you leave part of your manned spacecraft in lunar orbit and don't soft-land all of it on the Moon so you have to carry it up from the lunar surface again, you can save take-off weight on Earth. That was basic and challenged by nobody. But the question was, if the price you pay for that capability is one extra pressurized crew compartment, complete with life support equipment, two additional guidance systems to the one already in the command module, and the electrical power supply for all that gear, if you add up all this, will you still be on the plus side of your trade-offs? That was the real issue.

Some reporters told pretty fanciful tales that Houbolt had come up with an ingenious solution that everybody in Washington, Houston, and Marshall had overlooked. This is not true either, because special landing vehicles for the descent from and reascent to orbit of a target planet had been described in the literature before.

. . . It has been presented in some places as though this [von Braun's decision to switch to LOR] came as a great shock to everyone . . . including your own people.

It is correct that quite a few people at Marshall, who had worked on the Earth rendezvous system and convinced themselves that it was feasible, were disappointed when I sided with the lunar rendezvous mode that had been studied by Houston. But I would like to repeat once more that when I committed Marshall earlier to study the Earth rendezvous mode, I was not at that time saying that this was the mode we were going to embrace. At that stage, neither Bob Gilruth [director of the Manned Spacecraft Center in Houston] nor I were sure which way we should go. We had a charter to go to the Moon, and there were several feasible approaches and we simply agreed to study all of them. And we agreed that Marshall would analyze the Earth orbit rendezvous mode simply because we had already done a lot of work in that field in connection with space station studies. Houston would study the lunar orbit rendezvous mode and we would make a final selection only after both modes had been thoroughly wrung out.

Wernher von Braun, by this time the NASA deputy associate administrator for future programs, uses binoculars to monitor data on closed-circuit television screens in Firing Room 2 of the Launch Control Center during the final Apollo 14 launch preparations on January 31, 1971. (NASA Photo, available on NASA Image Exchange: http://www.nix.nasa.gov.)

I convinced myself after hearing the Houston story that in spite of the fact that the lunar orbit rendezvous story didn't look quite as gorgeous as it was originally presented by John Houbolt, it still looked like the best choice. By the way, Max Faget, Houston's key systems man, was probably the most vocal with respect to the inadequacies of the original Houbolt proposal.

"Your figures lie," he said.

I don't know whether Max really put it that bluntly, but I am sure I never did. John Houbolt is a very capable and dedicated man, and the last thing I wanted to do then and now is run him into the ground. But when he presented his original story to Houston, Max Faget was pretty outspoken.

The discrepancy between the original Houbolt proposal and the real world is best shown by two figures. Houbolt's fully loaded and fueled LM was to weigh a little less than 10,000 pounds. By comparison, the Apollo 11 LM actually weighed over 30,000 pounds. That is a flat 20,000 pound difference in the payload the Saturn V had to inject into a lunar trajectory!

While the two studies were underway, we at Marshall were fully aware of the fact that before Houston and Marshall could really compare the pros and cons of the two modes they had to be put on a comparable level of realism. Well, when the studies had finally been completed, I came to the conclusion that lunar orbit rendezvous still looked awfully good, in spite of the tremendous weight increases over the original Houbolt proposal, even after it had been brought up to the same degree of realism that we thought our own study had. This, of course, caused some disappointment on the part of some of our guys who had thought that Earth orbit rendezvous would come out on top.

We've spent a lot of time on this, but it is an important decision because your swinging over to LOR [lunar orbit rendezvous] was the most important factor.

Yes; I still don't like the term, "swinging over to LOR." I had never committed myself to EOR [Earth Orbit Rendezvous] in the first place. I've always taken the position, to repeat this again, that we in Marshall would investigate EOR and Houston would investigate LOR and that we would make a final decision on the mode after all the facts had been assembled. For that reason I never considered in any way that we were changing sides or anything like that. I just wasn't ready to vote at all until I had the facts and could make a meaningful comparison. Of course, some of our people at Marshall in the meantime fell in love with the scheme they were investigating. I guess that is only human.

Interview #2

Would you recall for us your recollection of the ARPA [Advanced Research Projects Agency] request for a clustered-engine booster and the evolution of the Saturn I.

The head of ARPA at that time was Roy Johnson. He was visiting the ABMA [Army Ballistic Missile Agency] at that time and said something to the effect that there was an indication that the Russians are working on a very powerful rocket system with a total thrust far exceeding anything we had, even in our ICBMs, the Atlas, and Titan II at that time, and there was also a widely held belief that they got this massive thrust by clustering a great number of engines. Would we be ready at ABMA to develop a powerful booster using existing rocket engines and clustering them?

Prior to that, back in about 1957, wasn't the future projects office working on this kind of clustering concept both with clustered H-1s and Redstone and Jupiter tanks, and with the parallel-stage concept which they were both developing?

Yes. But this was, I believe, already in response to an interest expressed by ARPA. I don't know whether we came forth with drawings of clustering rockets, or whether ARPA came to us. At any rate, it was pretty obvious that at ARPA there was interest in

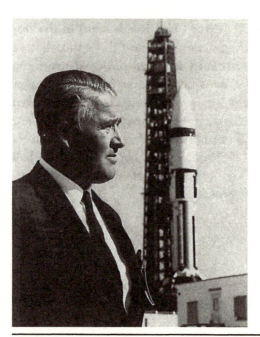

Wernher von Braun stands in front of a Saturn IB Launch Vehicle at Kennedy Space Center (KSC) on January 22, 1968. Von Braun was Marshall's first center director (1960-1970). Under his leadership Marshall was responsible for the development of the Saturn rockets, the Skylab project, and getting the United States into Space and landing on the Moon with the Apollo missions. (NASA Photo, available on NASA Image Exchange: *http://www.nix.nasa.gov.*)

high thrust systems. We were firm believers in the feasibility of clusters, and the question who made the opening statement is a little bit like who started a love affair . . .

Now there was another engine under discussion that was, however, only a paper engine, and I think it was called the E-1 which was halfway between the existing H-1 engine that came out of the Thor, and the F-1 which ultimately wound up in the Saturn V program.

. . . And so the early cluster studies involved a cluster of pure E-1 engines. But it was pretty obvious that we were probably more than $100,000,000 away from an E-1 engine in the funding environment that existed. This kind of money was beyond reach. So we ultimately settled on the idea of clustering eight H-1 engines to get that one and a half million pound thrust.

Going back a little further . . . several people have mentioned various paper studies made at Fort Bliss for large rocket motors using some of the Peenemunde technology and extrapolating . . . Was any application in mind for those large engines?

No. I think the question ARPA asked was really much more down-to-earth. They said something like, "We have $10 million, or so, for this purpose. Do you think this is enough to get us going? And in this environment, anything involving development of nonexisiting engines was simply not in the cards.

But there was some effort at Fort Bliss along these lines, wasn't there?

That is correct. But these were always paper studies whereas the H-1 was really in existence.

Were large boosters the ultimate objective of the Fort Bliss studies or were any space programs considered?

We made, of course, studies in Fort Bliss on the space potential of rockets. And since all those rockets turned out to be pretty big rockets, the question was, "How do you get the thrust?" And the thrust was, of course, in clustering a number of smaller engines . . . it's a bit like airplanes. If you have an engine of limited horsepower and you want to build a big airplane, people have always reverted to the technique of saying, "Oh, gee, let's take six or eight of these engines. If you put enough on, it will ultimately take off."

When you got the initial order to test the clustering concept and then about three months later you got an order to make this a flyable booster, was any time lost in this interim? Would you have designed it differently from the start, or done a different program if you had been asked to design a flyable booster from the start or did you kind of know what it [the cluster] was going to fly when you put it together?

We had in mind right from the beginning to come up with a configuration that would inherently fly. Remember, here again we were governed by low funding aspects, so we said, "How can we put a cluster of tanks together on top of this cluster of engines to provide the fuel, the RP [kerosene] and the LOX [liquid oxygen]?"

And we had, at that time, tooling both from the Redstone and from the Jupiter . . . And we made some very simple-minded sketches, figuring out how one could fit these two diameters together and wind up just with the right volumetric tankage to satisfy the needs of the engine.

Weren't the first studies modular in that the tanks could actually be plugged together, or unplugged, and shipped to the Cape and replugged again, before the barge concept came along?

That was discussed as one possible advantage of clustering tanks. But when you look at all the check-out requirements of the complete system, you want to verify and so forth. We all came to the conclusion that it's probably a lot easier to do a good job in the shop and make good pressure tests, seal the unit and ship it in one piece, rather than take it apart and reassemble it at the launch site.

Interview #1

The committee suggested that a fifth engine be included in the Saturn V. I always thought that this was your idea. Didn't you say at one time, "It's crying out for another engine?"

Yes, I have always pleaded for it, but this doesn't rule out that Milt Rosen's committee may also have recommended it . . . The term C-5, which is used throughout for the Saturn V, shows that at the time when the mode studies were made—I mean this comparison between Earth orbit rendezvous and lunar orbit rendezvous—we were already thinking seriously of a five-engine Saturn V. The term C-5 implies five F-1 engines in the first stage. The four-engine version was called C-4, and C-8, which you find occasionally referred to, had eight F-1 engines in the first stage and could have carried the Command Module to the lunar surface and back, so no LM would have been necessary. So, the LOR vs. EOR mode studies were based on five engines in the first stage . . .

We actually started with a launch vehicle with two F-1 engines in the first stage, then we studied one with four engines. And I figured, since all the performance and weight figures for the "front end" of the Saturn-Apollo vehicle were still extremely fluid—there were so many uncertainties about the LM, the weight of the command and service modules—that it would be wiser to provide some padding in the performance in the booster. I said relatively early that to build a Saturn with four engines in the first stage doesn't make sense to me! This great big hole in the center is crying for a fifth engine. And I was glad that this position was also taken by Milt Rosen's committee. Whether they came out with this suggestion before or after I did, I do not recollect . . . I would also say in retrospect, had it not been for that fifth engine, we would have been in deep trouble. Because the weight growth in the front end continued for two years.

Interview #2

When did the lunar rover come in as an added payload factor?

The rover is not exactly a new idea, you know. People had been toying with the idea of providing ground transportation on the Moon for a long time . . .

At the beginning we had very little information on the suitability of the lunar soil to accommodate wheeled vehicles. I remember that even after the successful landing of Apollo 11, I talked with Buzz Aldrin about this. And he said, "Well, it's going to be pretty rough driving around on the lunar surface since there are lots of craters there. I'm not saying you can't do it, but it's going to be a pretty rough ride." But the bearing strength of the lunar soil was there.

In subsequent flights, it became more obvious when the astronauts came back and said, "We just have too little time on the Moon to do all of the wonderful things we could have done. It would vastly increase the scientific payoff had we had a little more time and a little more facility to move around freely. We were just too damn busy doing—staying alive there."

After the first two landings on the Moon—Apollo 11 and Apollo 12—the question arose, "All right, we have not done even more than President Kennedy had promised in landing men on the Moon in the '70s. Here we still have a whole fleet of unused Saturn V Apollo systems and we have a substantial cadre of astronauts, competent, trained and eager to go. Shall we continue flying to the Moon?"

The scientific community, of course, had been at first a little skeptical about the scientific value of Apollo in the sense of, "Will the geological findings on the Moon really be so interesting that we ought to send man there?" . . .

After Apollo 11 and 12, and looking at the samples returned, they changed their minds completely. They came to the conclusion that the Moon really turned out to be sort of a Rosetta Stone for the understanding of monumental cosmological phenomena. So they were, all of a sudden, pushing. And they said, "Of course, if you want to continue to the Moon, stop flying to the mares because they are not the most interesting things on the Moon. Go to the more exciting sites . . . And that, of course, brought in the question of flexibility of the movability radius of action and increased payload capability of the Saturn V.

That fifth engine in the Saturn V still allowed you to stick the rover in?

Without the fifth engine, it would have been out of the question. They came to us and said, "Can you extend the system? Everybody complains about having too little time on the Moon; everybody complains about not having enough mobility on the Moon. Can you provide us with that?"

Of course, with that fifth engine, we still had a comfortable padding with respect to rocket performance, and so we could now do it.

Although, let me say this: Even with the fifth engine, we could accommodate the larger LEM descent stage that was necessary to soft-land the rover by further jazzing up the F-1 engines. We had to increase both the thrust and the specific impulse, in spite of the fact that we had five engines. So it was necessary to soup up the engines in addition to having five. With only four it would have been absolutely out of the question.

Interview #1

What do you remember about the C-prime decision—the Apollo 8 decision. I'll tell you, to orient you chronologically, it was on June 9, 1968, that Bob Gilruth called you from Houston and said, "We'd like to come over and talk." It was only that morning that George Low had first suggested flying an orbit around the Moon, and both George Mueller and Jim Webb had just taken off—

For Vienna—

For Vienna. And, as George Low says, or as Tom Paine said the other day, "It must have seemed a bit peculiar to them that the LM lost two months immediately when they got on the airplane." What do you remember about this meeting in Huntsville on that afternoon?

Bob Gilruth's story was really very simple. The original plan, you remember, was to fly Apollo 8 as the first LM mission, pretty much as it later was flown on Apollo 9, in Earth orbit. Only thereafter were we to fly a lunar mission, pretty much on the profile of what became Apollo 10, the Stafford mission. Now Gilruth said, "No matter what we decide to do, that LM just isn't ready to fly. And, so if you Marshall guys say our Saturn V is only ready to launch this Earth orbit mission for a command module/LM rendezvous in Earth orbit, all we can tell you is we will have to wait three months. However, if we wanted to fly Apollo 8 as a mission with the CSM only, and without the LM, we could crank up pretty soon. And if we did that, we might as well swing around the Moon. Well, our position was that the Saturn V really didn't care whether it headed for Earth orbit or on to the Moon. We had reignited the SIV-B stage in orbit before. As far as the other potential problems with the launch vehicle were concerned, we thought we had smoked out the problems we had in Apollo 6. Therefore, we would be comfortable with a decision to go around the Moon with Apollo 8. If I remember correctly, in his opening remarks Gilruth talked only of flying around the Moon, not of going into orbit around the Moon.

After the Houston fellows had given us their basic pitch, we debated at length the wisdom of just swinging around the Moon through a free-return trajectory versus going into orbit around the Moon. The Houston group felt the CSM was ready to go into orbit around the Moon as well. They felt they would learn a lot about tracking the spacecraft out there and really nailing down the entire flight profile of future Apollo flights, so it would really be worthwhile. There would be a tremendous gain in time if that portion of future lunar flights could already be exercised with Mission Control.

I still think the boldest part of the decision was to commit the launch vehicle just after you had had so much trouble with it on the previous flight [Apollo 6].

We didn't feel too concerned about the launch vehicle risk because we really felt we were on top of the Apollo 6 problems. Also, there is a built-in emergency return capability. For example, if we tried to inject the Apollo 8 spacecraft all the way to the Moon and 10 seconds before nominal cut-off the SIV-B would have acted up, there would be enough Delta-Vee left in the service module to bring the crew safely home. We had practically all the built-in abort options provided for the Apollo program available to Apollo 8. Of course, there was no operational experience with this sort of flight profile but the Delta-Vee reserve for contingencies was there. From our point of view, the risk difference between a Saturn V launch to Earth orbit and to go from there on to the Moon was a relatively small thing.

We had a very good session there in Huntsville, and we thought the Houston people had a good and convincing story: The LM wasn't ready. We were up against the choice of waiting at least until February '69 to fly Apollo 8 through what later became the

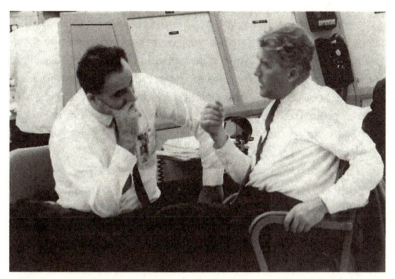

Wernher von Braun and Rocco Petrone, director of launch operations, talk during a lull in the preparations of a Saturn I vehicle launch at Cape Kennedy's Launch complex 37 Control Center on May 25, 1965. (NASA Photo, available on NASA Image Exchange: *http://www.nix.nasa.gov*.)

Apollo 9 profile, or flying Apollo 8 without LM around the Moon. Of course, after we at Marshall had endorsed Houston's plan, the much more difficult problem was to convince our doubting bosses in far-away Vienna. Jim Webb and George Mueller must have figured that no sooner had they left Washington that we were trying to tell them by remote control how to run NASA.

Interview #2

Could you compare Russian boosters and their concept with the American system, the NASA system, and make some comments as to why the Saturn hardware was more successful, or why we got to the Moon first ...

This is a very involved question. I think that the Russian launch vehicles have already proven to be quite reliable ...

There is probably less competition in the Soviet launch vehicle program than in this country. The relationship between the space people, the space program people, and the Soviet Union, and the rocket people, is probably best compared with the relationship we have between these two groups in this country during the Gemini program where, as you will remember, NASA built spacecraft but went to the Air Force to request Titan II launch services for Gemini spacecraft.

The launching itself of Gemini spacecraft was done largely by blue-suiters. And it was only in the Apollo program that we brought a launch vehicle into the process that had no military history at all. Remember, even the Mercury used Atlas launch vehicles, and the Redstone rocket preceded Atlas very early— Alan Shepard's and Gus Grissom's flights had a military history.

The Saturn V was really the first launch rocket that was a baby of NASA and not the military—a military child.

Now the entire family of Soviet launch vehicles up to this point was really developed under military auspices. They have the so-called Strategic Rocket Command in the Soviet Union, comparable to our Strategic Air Command, and they are really the sole owners of rocketry, you might say. And the space people go to them for booster service, just like NASA went to the Air Force for Atlases and Geminis.

The industrial complex—if that's what you want to call it— state-controlled economy—undoubtedly doesn't have as many facets as the American aerospace industry. In other words, they don't have their Boeings, and North American Rockwells, and Douglases, and so forth, to build competing systems. But it was, and I believe still is, a more monolithic operation.

With that I am not saying there's no competition at all. I think there's every indication that within that monolithic industrial structure there are some competing teams. You see that in their aviation industry . . .

Nevertheless, I think it is far more monolithic—and that also means that, shall we say, there are less checks and balances in this. In NASA, you could always tell the Boeing people, "Look, the Douglas people brought something in here which, in our opinion, greatly enhanced the liability of something," and vice versa. So the government was in the fortunate position that it could effectively cross-feed ideas that came out of these various pots.

When you have a very monolithic organization, that is one, shall we say, like a military establishment, you have less and less of that. There is, at the end, one man responsible for all these things. You know, the Russians always mysteriously refer to "the chief constructor," or "the chief engineer," whoever that man is . . .

We have never run the Saturn V program like that in NASA. I think we considered ourselves far more like a stock exchange of good ideas where we felt we picked the best things out of all these things and cross-fed them for maximum benefit of the whole.

Dr. Robert R. Gilruth, director, Manned Spacecraft Center (renamed the Johnson Space Center in 1973), Houston, Texas. Photo was taken in 1965. (NASA Photo S-65-45100.)

ROBERT R. GILRUTH

(1913 – 2000)

Dr. Robert Rowe Gilruth's entire professional career has been devoted to government service, first with the National Advisory Committee for Aeronautics (NACA) and then with its successor, NASA. In his early years, Dr. Gilruth was deeply involved in transonic aerodynamic research. He used free-falling instrumented bodies dropped from high altitudes, and the wing-flow technique he had invented, to solve problems of aerodynamics near and through the speed of sound.

Born on October 8, 1913, in Nashwauk, Minnesota, Dr. Gilruth received a bachelor of science degree in aeronautical engineering in 1935 from the University of Minnesota, followed by a master of science in aeronautical engineering in 1936.

In 1945, the United States Congress approved the establishment of a free-flight guided missile range to be built at Wallops Island, Virginia, and operated by NACA. Dr. Gilruth was selected to manage this new facility. Although he was a young man of 31 with little experience except research, he soon mastered this new world of budgets, land acquisition, recruiting, and operating with other government agencies and industry. Originally conceived as a missile test site, Wallops soon became a missile research range and Dr. Gilruth began developing basic information on the aerodynamic and structural behavior of wings, bodies, controls, and other key items in missile and aircraft design.

In 1952, Dr. Gilruth was appointed assistant director of the Langley Laboratory when, during this period, he acquired a nucleus of people who were skillful and creative in research with rocket-powered models in free-flight at supersonic and transonic speeds. He directed research efforts in hypersonic aerodynamics at the Wallops Island Station and the research in high-temperature structures and dynamic loads at the Langley Laboratory. The results of Dr. Gilruth's research on the behavior of wings and controls at sonic speeds, and his analysis of wing flow tests on thin wing sections for aircraft designed to fly men through the speed of sound, were available in time to be considered in the very thin wing section for the X-1 research aircraft, the first aircraft to "break the sound barrier." Because of the Wallops Island research results, there were practically no surprises during this historic flight. When the X-15 aircraft, which had been proposed for flying men to the very edge of space and back, became a successful research aircraft, Dr. Gilruth and his associates began thinking about manned satellites.

With the replacement of NACA with NASA in 1958, Dr. Gilruth became the director of the Space Task Group at Langley Field, Virginia, the organization responsible for the design, development, and flight operations of Project Mercury, the nation's first manned space flight program. It was during this time that Dr. Gilruth helped organize the Manned Spacecraft Center (now the Johnson Space Center) in Houston, Texas, and selected a highly competent workforce capable of performing the many diverse functions required for a program of this magnitude.

Leading that effort, Dr. Gilruth became director of the Manned Spacecraft Center in 1961 where he served until January 1972 when he took on a new position within NASA as director of key personnel development. In this capacity, he was responsible for identifying near- and long-range potential candidates for key jobs in the agency and for creating plans and procedures that would aid in the development of these candidates.

In December 1973, Dr. Gilruth retired from NASA and, in January 1974, was appointed as a consultant to the administrator. In February of that same year, Dr. Gilruth was appointed to the Board of Directors of Bunker Ramo Corporation. He was also appointed a *member of the National Academy of Engineering (NAE) Aeronautics and Space Engineering Board (ASEB), and asked to serve as a member of the Houston Chamber of Commerce Energy Task Force.*

During his long and distinguished civil service career, Dr. Gilruth has received many honors including being elected a member of the National Academy of Sciences and a member of the National Academy of Engineering. In 1972, he received the Robert J. Collier Trophy for the greatest achievement in aeronautics or astronautics. He is also an honorary fellow in the American Institute of Aeronautics and Astronautics; a fellow in the American Astronautical Society; an honorary fellow of the Royal Aeronautical Society; and a member of the International Academy of Astronautics. In addition, Dr. Gilruth served on numerous scientific advisory committees for the military services and NASA. Honorary degrees bestowed upon him include honorary degrees of doctor of science from the University of Minnesota, the Indiana Institute of Technology, and the George Washington University. He received an honorary degree of doctor of engineering from Michigan Technological University, and an honorary doctor of law degree from New Mexico State University.

_____

Editor's note: The following are edited excerpts from four separate interviews conducted with Robert R. Gilruth. Robert Sherrod conducted interview #1 on November 16, 1972. David DeVorkin and John Mauer conducted interview #2 on October 2, 1986. David DeVorkin and John Mauer conducted interview #3 on February 28, 1987. David DeVorkin and Martin Collins conducted interview #4 on March 2, 1987. The last three interviews were conducted as part of the Smithsonian Institute's National Air and Space Museum Glenn-Webb-Seamans Project for Research in Space History.[1]

Interview #1

The Soviets, having launched Sputnik, in a way, made putting a man into space worthwhile. You had doubts about it before, as to whether it was just a stunt or not, but this now gave it meaning. I don't want to put words into your mouth, and if I'm not saying it correctly, indicate what your opinion was at that time.

Let me tell you that to the best of my knowledge, I never thought of flying people in space before Sputnik. When Sputnik went up, it was a shock. And it was not just a shock to me, but it was a shock all the way through the technical people of the United States, including a man who was my "big boss" in Washington. By that time, Dr. George Lewis, who had been the head of NACA for many years and was responsible for all the wind tunnels, and Dr. Hugh Dryden, who was affiliated with the National Bureau of Standards, were working sort of free-lance at the time after Sputnik went up, trying to help the country get hold of itself and get a program that would do well. Sputnik went up on October 4, 1957, and on November 3 they put a second Sputnik up with a dog [named Laika] in it. On that second flight I said to myself and to my colleagues, "This means that the Soviets are going to fly a man in space." They've got so many accolades from all of the people all around the world. The Soviet stock, so to speak, went way up, and they began to emerge as the leading technical nation in the world.

So after the dog went up—that was on November 3— on November 7 President Dwight D. Eisenhower appointed Dr. James R. Killian as special assistant to the president for science and technology, which meant that even Eisenhower was affected quite heavily by this. Then, on November 21, Dr. Dryden, who was filling in now for NACA, created a Space Technology Committee at NACA. He was, of course, on that Committee along with Dr. Wernher von Braun, Guy Stever—who was a member of many committees, a good man, and myself. Also, Jim Dempsey, who was head of the firm that was building the Atlas rocket, and Dr. Randolph Lovelace, who was the head of the Lovelace Institute, a doctor who was a very big help to us in our man-in-space programs. There were several other people as well. So there were quite rapid reactions to the fact that Sputnik went up. Jimmy Doolittle was also very instrumental along with Dryden in talking to the President mostly through the President's Science Advisory Committee. That was pretty much the situation. Early in 1958, the people at Langley worked on a space document for Headquarters that described the activities that we already had going in the space field which showed that NACA was active in this area.

Was it on November 5, 1958, that the Space Task Group, originally called the Task Group, was actually officially formed?

Yes. I saw that date somewhere. We called ourselves the Space Task Group before anybody knew why we did it.

Do you remember where the term "Space Task Group" came from?

I think it was the best thing I could think of. A lot of people wanted to call it the Task Force. I didn't think that was a good thing, Task Force. It was a group.

Not a Tiger Team?

Not a Tiger Team, not a Skunk Works. It wasn't any of those things; it was a Task Group.

I'd like to get at just that question, following that up. Why is group better than force for you?

Because force is an adjective that implies you have a lot of strength. I didn't know whether we had a lot of strength or not.

Some people, when they build, try to appear as if they have the strength, hoping that this will make a lot of people assume they have it. They basically assume the position, and what they expect. Others build from the inside.

We had the strength. We had the money. What we needed was a good name, and we didn't want to sound like we thought we were too big for our britches. We were a Task Group. We had a big task and we were a group. It was a very good name. I wouldn't have changed it for Task Force. A Task Force is something you use in the military.

Interview #2

Let's talk a little bit about the first few months, the first year of the transition [from NACA to NASA]. I'd like to know—as you were asked to be on these task forces—were you given any kind of a choice and did you consider any different moves? All of Langley apparently was being absorbed, and I guess you knew that. Is that correct?

No. No, Langley wasn't. I was pried out of Langley and I started the Space Task Group. And I was sent back to Langley. I wrote to my friend, Tommy [Floyd L.] Thompson [director of the Langley Research Center], and said "I've been authorized by the administrator to draw out certain people from your staff and these are the ones I'd like." I listed a page and a half of key people that I wanted, and he let me have everyone except one. That was when one of his division chiefs said, "Bob [Gilruth] doesn't really need that one, we'll give you somebody else," but that one I really needed because he was doing a special job for me.

But did you have the feeling that you were being pulled out of something that you would prefer doing? Was this something that you had to think twice about?

I didn't think I wanted to do this space business all my life, but I was fascinated by it. I thought it was terribly dangerous and probably I'd end up in jail or something, but I really thought it was important to do and I was having a lot of fun.

So you didn't resist it and you didn't see this as a dangerous move in your career?

I thought it was probably a high-risk move. If I failed, I knew that would be very bad. On the other hand, if it worked maybe that would be good.

Did you know what was expected of you?

Yes, I was expected to put man in space and bring him back in good shape—and do it before the Soviets, which we didn't do. On the other hand, we buried the Soviets before we were through by going to the Moon.

Was there a sense of failure because the Soviets constantly—in the initial few years—were ahead?

No, because we had no chance of overcoming them that fast. They didn't even say they had a program to put man in space. They never admitted they had a program to put man in space until they put the gear in space.

But you knew once Laika went up?

Once Laika . . . I said they've got to be doing it, sure.

One of my questions was, in reading [an earlier interview] where you talk a lot about the popular media's conception of the effect of

Sputnik, that it was a terrible surprise, that we were behind, that this was a threat to national security. You seem to think that it wasn't so much a threat until you realized what they could really do with some of the later Sputniks; Sputnik I didn't impress you that much. But, I'm wondering, did you see it as a threat or as a real opportunity to go from having, in the early mode of activity, a plan without a program, to a real program?

I think that as this thing unfolded, it wasn't going to be just one flight of man as the whole program. And that's why when Kennedy came along and said, "Look, I want to be first. Now do something." I said, "Well, you've got to pick a job that's so difficult, that it's new, that they'll have to start from scratch. They just can't take their old rocket and put another gimmick on it and do something we can't do. It's got to be something that requires a great big rocket, like going to the Moon. Going to the Moon will take new rockets, new technology, and if you want to do that, I think our country could probably win because we'd both have to start from scratch."

I usually heard this kind of logic attributed to von Braun. Is that correct? Or were there many people who had that logic?

Well, I think Wernher had that, too, but I was the one who was talking to Kennedy.

You were talking to him personally?

Yes . . . I was talking to him. And I told him that very thing. If you really want to be first, you've got to take something that is so difficult we'll both have to start from scratch.

You're speaking of Kennedy . . . Did you feel that you had a real program, or was it still a career risk for you during the Glennan years?

Well, I was not really happy with Glennan and I wasn't very happy with Eisenhower, because Eisenhower obviously didn't believe in the program. He'd been talked into it and he was reluctant. You know, he took money away from the program in the last few months of his term. I'd seen Eisenhower, but I never met him or talked to him about it.

But you did meet Kennedy?

I saw Kennedy many times.

You state very clearly that the strength of the NACA was in the centers. And now you're moving into an area where there was going to be contracting. NASA was going to be a contracting organization, but it was going to bring in the NACA centers, it was going to be built on the NACA. Were you worried?

What do you mean, "built on the NACA?"

NASA in its early design was to be based upon the NACA centers. Is that a fair statement? Langley was going to be brought in, Ames, Lewis . . .

Yes, they were all going to be part of NASA.

Previously, there had been research centers either responding or dreaming up new research. Dryden had said or had hoped that NASA would retain these centers, these old NACA Centers, as research centers. Were you in fear that they would be changed from being pure research centers to becoming simply centers for contracting work out? A few minutes ago we said that you couldn't produce a new center that could put a man on the Moon. You had the contract with industry and you became literally a contracting agency. Were you worried that the vitality, the research vitality of the centers would be affected?

The way it worked . . .Langley, which was a research center, was very proud of the fact that it was a research center, and they did not want to become a contracting center. We who had a job of putting a man in space knew damn well that we had the luxury of having a few good men in a few laboratories that we could do some work in support of our programs, so we would not be entirely dependent on the other research centers. And because we always like to have people who liked to do that kind of work, just to talk to . . .

You had to remain technically vital.

Yes. And so when we built our center—thanks to Mr. Kennedy and the Moon program—we were sent down to Houston where we built a center. Believe me, we built it from scratch. There was nothing there but a field with some cows on it.

But you didn't know this until Kennedy came in.

That's right. We were responding to a challenge of the country. We weren't sure what was going to happen. We thought we could do that job and if it turned out to be a dead-end job, the fact that we'd been able to do it, we felt, would allow us to do jobs elsewhere. I don't think we had a very big risk except of not being able to pull it off. That was a big risk.

Interview #3

In August 1959, you created what was called the New Projects Panel to identify new areas for research. I'm interested in how it was set up. You made [Kurt] Strauss the head.

Yes, Kurt Strauss. Bob Piland worked on that, too . . . We came up with this. We thought we ought to be looking ahead at what kind of a spacecraft we were looking at next? We came up with a place where you could do work, where you could do tests of weightlessness. We came up with a bigger capsule than Mercury, along with a tank-like object in which you had room to do experiments. That was really the forerunner of the lunar module. We worked hard on this thing, then when we came along with the lunar program, we were able to modify some of this and add parts to it. We actually saved a little time by having been looking ahead at some of these things.

Key leaders in the Gemini program are shown during a press conference at the Manned Spacecraft Center in Houston on May 17, 1966, following the Gemini IX Atlas Agena failure when one of the two Atlas outboard engines gimbaled and locked into a hardover pitchdown position. As a result, the whole combination—Atlas and Agena—flipped over into a nosedive plunging the vehicle into the Atlantic Ocean, 198 kilometers from where it was launched. Shown left to right in this photo are: Charles Berry, director of Medical Research and Operations, MSC; Deke Slayton, Director of Flight Crew Operations, MSC; Gene Kranz, Flight Director for Gemini IX, MSC; Charles Mathews, Gemini program manager, NASA HQ; William Schneider, Director of Mission Operations, NASA HQ; Lt. General Leighton Davis, Department of Defense Liaison; Dr. Robert Gilruth, Director, MSC; and Dr. George Mueller, Associate Administrator for Manned Space Flight, NASA HQ. (NASA Photo 66-32629.)

When did you start thinking that there was going to be a center for manned spacecraft?

It became obvious. Mercury was a dead-end program. You were going to fly man in space, orbit him—that was it. I was ready to go back and sail my boat after that. Anyway, it didn't work out that way. We ended up with Kennedy and going to the Moon. That was a big program. There was no question that it couldn't be handled out of Langley, and it couldn't have been handled out of Washington. It had to be handled out of Texas, because that's where the head of the Appropriations Committee lived. Right smack there in Houston.

So the Mercury program itself didn't warrant thinking really big?

That's right. It was a dead-end program.

You were aware of that right along?

It seemed to me that it was because you never stop with the one thing you're doing. But it wasn't obvious what the next step was.

What was it about your experiences that let you have confidence in getting into this whole new area of dealing with contracts?

I didn't have time to think about that problem. It seemed to be very straightforward. The people at McDonnell, I'd known before. I knew Mr. McDonnell and [John F.] Yardley, a stress analysis man who late became president.

So you already knew many of the key contractors?

I knew them. But I'd never had a relationship where they were the contractors. I was "Bob" and we could talk about what the problems were. We really didn't have any bad problems. We had arguments about how the capsule should be built. And we resolved those on a rational basis, and I think we were satisfied. It wasn't hard to do that. The Atlas was a little different because we were buying it from General Dynamics through the Air Force. Of course the Air Force was really quite concerned that we might give their rocket a bad name because they weren't sure that we knew what we were doing. And especially when we—the second Atlas we flew blew up at Mach 1 just 60 seconds after it was started, and—

But that was prepared by the Air Force for you, wasn't it?

Yes. But see, it was a case of, we had our spacecraft on the front, and they said, "Well, that came apart and went back and hit the Atlas and caused it to blow." Which it might have.

They also had motivation to blame your spacecraft, because if they blamed your spacecraft, then their rocket wasn't to blame.

Sure. And obviously, there was a case for an argument, and we wanted to fly again. And I said, "I believe that the skin of your rocket is too thin," I think it was 20/1000th. The Atlas was a stainless steel balloon. I said, "The turbulence from our spacecraft is probably enough to cause that to oscillate and wrinkle and it probably ruptures." So I wanted to put a collar around it of heavier stainless steel with bands that would tighten it up, and I made the mistake of calling it a "belly band." This was very, very unpopular with the Air Force. But I said, "We paid for this rocket, and we need to fly." They said, "Well, we'll put heavier gauge material on the rocket, but that will take five months to do." I'd already looked into that, and I said, "We can't wait five months. We need to put the belly band on." Nobody in the Air Force would approve it. We finally carried it all the way to the Secretary of the Air Force, who I'd known. I said, "I will agree to take all blame if it breaks," which meant I would be out of a job. I said I was the guy that insisted on doing this—I said, "okay, I'll take the blame if it blows." Well, we put the belly band on, and I remember very well when we launched that thing. I went outside all by myself, behind a bush there, and watched that thing go. I kept timing it, and when 60 seconds elapsed, that was the time it went through Mach 1. It kept staying together. Then I went back into the blockhouse and I said, "Well, we can relax a little bit. We didn't have that problem again." That saved us four or five months in our program.

You came up with the idea, or someone on your staff, and took responsibility for it, the belly band?

The belly band? Oh, I don't know. I think maybe it was somebody like Yardley who said, "Well, what we need to do is to strengthen that up," and so we got the idea of the belly band. But it was my idea to carry the thing through the hard knocks.

You never asked somebody at Headquarters to vouch for you?

Well, Jim Webb was head of NASA and he knew all about it. I told him. He knew our problem.

But you were the one who was ready to take the heat.

Yes, somebody had to take the heat and I was the logical one. And maybe he [Jim Webb] would have saved my neck, if it had blown up, but I don't think he could have. In any case, that's a true story. That was really an important thing, because we couldn't have stood that four or five month delay.

Let me ask, why did you leave the bunker? Why did you go outside to watch it?

Because I wanted to get the best possible view of what was happening. Inside, you couldn't see very much. You could see all the instruments and everything.

But it sounds like, having put your job on the line, you were also willing to put your life on the line.

Oh, no, no. It wasn't dangerous.

It wasn't, even at launch?

I wasn't that close to it.

In listening to your account, it strikes me differently, maybe it's in the record and I just didn't pick up on it before—but the failure that prompted you to come to the belly band decision, was it directly associated with the dynamic forces of going through the sound barrier at Mach 1?

I think so.

And is that why you went outside, because you wanted to see the rocket's actual performance when it went through the sound barrier?

Well, you could see the failure better from outside than you could on a television screen inside.

If that problem was going to be repeated was it going to be repeated at Mach 1?

Yes. That's where it happened before, and that's where the buffeting was the worst. If you got through that, I thought we were perfectly all right. It turned out that it was. From then on, we had three good Atlases in a row. No, four. I think every one. Glenn and all the other astronauts except Grissom and Shepard flew on that Atlas—Deke Slayton, of course, didn't fly.

Did they fly Atlases with belly bands?

No, from then on they had the thick skin.

So the belly band was a temporary thing to bridge between the very thin one to the new one.

That's all it was, so you could make your flight without losing five months.

Did you ever have a feeling, during the Mercury Program, that without rules in the beginning, without an adequate infrastructure to monitor the contracts and all of the different phases of Mercury, especially at the point where you were also beginning to worry about building a new center, that things might get out of hand? Did you have contingencies, fallback positions?

I don't think that we felt particularly bad off in our monitoring of the Mercury spacecraft, for example. Or the Atlas rockets, because we were just buying those Atlas rockets, and we had one or two men there to investigate—when we were worried about the pumps. We routinely took the pumps apart to check the clearances in them, which was not a regular procedure.

But you did do that.

We did that. We had three or four good Atlases in a row, which is more than they usually got. And we didn't have very many people for some of those contracts. The price was already set by the Air Force. We just bought it through the Air Force.

So in a way, Mercury is different because you're just focussing upon specific parts of the program. Other parts, you're just buying in to what already was being developed.

That's right. There were some changes that were made. We wanted a different skin gauge on the front end of the Atlas. We had another rocket that we used, the Redstone, for suborbital flights. We were able to get them from von Braun and his people, and they were glad to be with us. They were a big help to us. We also had one of our own rockets. We had a solid rocket launcher that we used at Wallops Island to test some of the spacecraft. It was just a great big solid—

Little Joe?

Little Joe. And we used that for some of the escape system tests to save money.

In a way it's an advantage to buy a technology that's coming onto line because you don't have to worry about shepherding it through. On the other hand you're buying a technology that's blowing up on the pad a lot. And you're wanting to put a man on top of that—

That's right.

Did it make you nervous when you thought, we don't have control over the Atlas. We're buying what already exists. To some degree it did, but you give the example about checking the tolerances on the pumps.

There were some things we could do, and that was one of them— the tolerances on the pumps. We did increase the skin gauge. But the Atlas was an Atlas, and we knew that we were going to have to live with what existed.

Why did you feel that you could live with it? What was it about the Atlas that made you feel, this can work, even though there are some problems?

We had no alternative. We had to. And we did. We only had one Atlas blow. That was the first. We used an Atlas on Big Joe, and that was in the first year of our program. We sent a Mercury, made in our own shop, a Mercury capsule shape that we used to measure the heat transfer and stability. We sent that up and drove it back down into the atmosphere. It was a marvelous test. It showed that the space-craft was stable, that the heat transfer was what we thought it would be, and that a man could survive. Since we couldn't find the capsule right away, the papers all went to press and said it was another bust in the space program. The Navy found the capsule in a couple of hours and it was in good condition. The newspapers wouldn't pub-lish the fact that it was found . . .

It must have been a very poignant time. How serious was, to you personally, press attention, and how adequate was it in general in those first years?

Our first year it was a strange situation. The Air Force ran the mis-sile range down at the Banana River [Florida]. The headman was kind of like a little Napoleon. He was the boss and right after every launch he'd call all the press in. He'd get the people who had made the flight to stand up and say how it worked. In our case we didn't know how it worked because we were recovering something that was way down range. It wasn't just whether it blew up or not within your view. We couldn't have a good meeting because we didn't know how it worked. We hadn't yet recovered it. So the press went with the fact that it was lost . . .

In that situation, five months was at a premium because of the intensity of the race with the Soviets.

At that time, Gagarin hadn't flown yet. We were still hoping that we could somehow or other luck out and orbit a man first, but Gagarin flew right before we flew Al Shepard. The Wiesner Committee held us up—otherwise, at least, we could have—

The Wiesner Committee held you up?

Sure.

In what respect? I don't remember that part of the story.

When the Republicans lost and Kennedy took over, the Science Advisor became Jerry Wiesner of MIT, and he was very much against the man-in-space program, and Kennedy had no polarity at all. He didn't know much about it and was not interested. So Wiesner decided to hold hearings on the man-in-space program, with the idea that it should be stopped. But Jim Webb had also been picked by the Republicans. Jim Webb and I had fortunately gotten to know each other, and he thought that we had a good thing going. He wasn't supposed to administrate a program that lost all of its guts, so he was my friend. When Wiesner had these hearings to see whether or not Mercury should be cancelled—

This was in January of 1961 . . . early 1961, right after Kennedy became President?

Yes.

And [T. Keith] Glennan was—

78

Flight director Chris Kraft (left) and Dr. Robert Gilruth, MSC director, exchange congratulations in the Mission Control Center after learning that Astronauts L. Gordon Cooper Jr. and Charles "Pete" Conrad Jr. had been recovered in the western Atlantic to conclude the successful eight-day Gemini V mission on August 29, 1965. (NASA Photo S-65-44212.)

Glennan was no longer the Administrator of NASA.

That's right.

Webb was the new man running NASA. So the Wiesner Committee had some doctors on board who would not believe that man could stand weightlessness even for a few seconds. I said, "Well, we can take you up in an airplane and fly weightless, and it doesn't do anything bad to you. It hasn't hurt the monkeys that we orbited. We've already done that." Of course, we were going to make ballistic flights, up and down, like Al Shepard's flight. We hadn't made that yet. And we were trying to get permission to make the Al Shepard flight, and we couldn't get permission from Wiesner that it was safe, not from the point of view of the rocket blowing up, but from a point of view of, was it worth it and could you stand weightlessness? This went on for quite a while, and finally Webb got frustrated, and said that we're going to go ahead and fly. We believe that we've done everything we can do. We have this program and we're going to fly, and if you don't think that's right, you will have to make your case in the newspapers.

This is Webb talking to Wiesner?

I don't think he talked directly to Wiesner, but he gave him that message. And so that's the way it ended. We went ahead and flew Al Shepard, and when Kennedy saw how the American people loved that flight, it was all over as far as Wiesner was concerned.

Interview #3

Some people are nostalgic for that Mercury period, because of it being so much freer, people making a lot of decisions quickly. How do you feel? How does it compare to the time before, and how does it compare to after you were already at the Manned Spacecraft Center? Does it seem to be a better time or is it just another stage in your career? How do you view it?

I don't think you could live through many of these Mercury programs. It was something you do when you're young. You couldn't keep on doing that kind of thing. It was a case of working all the time, for the first year or so. But it was rewarding. It was great when Al Shepard flew, and when Glenn, and all the others flew—we were extremely fortunate to have all those things work.

It was the fact that it was so very successful, I believe, that we went on to the lunar program. Although it is true that Kennedy really got that going before we ever orbited John Glenn. I think the momentum of those Mercury flights had a lot to do with the success of the Apollo Program over those years, because it made it a lot easier to get the money that it took. It took a lot more money to build Apollo than it did Mercury. Apollo was $20 to $30 billion, and Mercury was, I think, closed out finally at about $400 million. We didn't think it would cost that much, but it did.

Interview #4

Why don't we go ahead and discuss the building of the Manned Spacecraft Center, later named the Johnson Space Center, in Houston, Texas. This was all part of the first year of decision, as you've described it, June 1961 through June 1962, when a center was found to be necessary.

Yes.

Where do we begin? I think we could use your advice on this. You were worrying about a hundred things at once. By then you had three major programs to concern yourself with: Mercury, which was operational, Gemini, and Apollo. You had manpower problems in simply determining who was going to work on which projects, where people could best be assigned, bringing new people in, worrying about everything from launch site to the goals and missions of the government, to the type of orbit, launch vehicles, test and operations, astronaut selection and training, orbit analysis, to celestial mechanics, just to name a few.

Yes. On top of that, we had to move.

But you didn't have to move.

Oh, yes we did.

Why?

In order to get the political backing we needed.

Okay, but you were in a program that was a hallmark of efficiency and directness and supposedly an absolute crash program to be done as efficiently as possible. **What sense did it make to you at the time, and to the people you worked with who were going to be affected, moving from Langley to anywhere else? Didn't it make more sense to stay?**

Yes, it did make more sense to stay. But I can quote you how Mr. [James] Webb handled that. He said, "We've got to get the power. We've got to get the money, or we can't do this program. And we've got to do it. And the first thing," he said, "we've got to move to Texas. Texas is a good place for you to operate. It's in the center of the country. You're on salt water. It happens also to be the home of the man who is the controller of the money." That was Albert Thomas.[2] So we moved. It turned out that it wasn't all that bad. We would have had to expand like hell anyway. When we went to Houston, we went into temporary buildings, which we would have had to do at Langley or at Hampton or at Newport News or somewhere. And it wasn't as good a place to live. The people weren't as gung-ho in Virginia as they were in Texas. They wanted us in Texas. They were thrilled to have the space program come to Texas.

Let's talk a bit about what he did do, and that is, take a person like John F. Parsons [associate director of the Ames Research Center] and send him out to do an open site survey of 10 possible sites in the United States, for where the Manned Spacecraft Center might be. Was this a necessary thing to do politically, or just a sham?

I think it was good to know what the reception would be in other parts of the country and what the advantages were. Actually, looking in retrospect at all the different places that were surveyed, I think they picked the best place.

Had you been there during the summer?

Yes, but it's not too nice in a lot of other places during different times of the year.

Some of the people that came down from Langley initially ended up going back to Langley, just because they found the jump too much. But for you personally, you didn't feel like the jump was that big a deal, in terms of the climate, the situation?

No. I thought it was a lousy climate. But the air conditioning was good, and the people were nice, and it had other things. The winters were quite good compared to what we have here in Virginia. So it wasn't all bad.

Many people have the assumption, so I would like to address it from that point of view, that it was Lyndon Baines Johnson who was one of the key driving forces for the selection of Texas and Houston. You are pointing to Albert Thomas as being much more the important political influence. But how much of a role was the fact that LBJ was from Texas?

I think he was important also. But I think Albert Thomas was more powerful.

It's the difference between being chair of the key committee in the House and being vice president. You're not that powerful when you're vice president.

LBJ was very anxious to have Texas get that, too. But I don't know. I was just the innocent person there. We went down and looked at Houston. Webb said, "Why don't you go down there and look?" I looked around, and I was happy to see there was salt water, and there was lots of open space. Around Clear Lake it was fine. We looked at where we could build the center, and there were cows munching on the grass. I still have a picture of that site, with cows roaming on it. When I go back and look at that center now, I'm very proud of what was done. It's beautiful and it's been very valuable to the United States. So I feel that maybe Albert Thomas had a different motive, but it worked out to be in the interest of the United States of America.

Donald Douglas Sr., Chairman of the Board of Douglas Aircraft Corporation, visits with Dr. Robert Gilruth (right) at MSC on August 11, 1965. (NASA Photo S-65-28500.)

Since there was a difference in the management approaches of [T. Keith] Glennan and [James] Webb, how did [Hugh] Dryden's role change between the time of working with Glennan from working with Webb?

I don't think his role changed much. He was the technical center. He was the last resort on any real problem that was technical. Neither Glennan nor Webb would try to vie with his knowledge of technology and science. It made a very good combination.

How did the strength that Dryden provided, in terms of his technical knowledge and his ability to give administrative leadership, change after he was no longer with NASA? Did it make any differences for you as a center director?

There really wasn't any way of replacing Dryden. Dryden was a big loss. Of course, it was near the end there. I can't remember exactly when Hugh died.

It was 1965, I believe.

It's before we landed on the Moon. He had cancer and he would get treatments. When he'd get a treatment he'd be feeling good for a while. Then after two or three weeks he'd begin to drag, and finally he'd go in for another treatment and then he'd feel good. But he said, "You know, this recurrence will not go on forever, and each time it's a little worse." He knew it was inevitable. A very brave wonderful man, highly religious man. I can remember it was during Gemini that he passed away. It was the first time I ever left mission control during a manned mission, when I went to his funeral. God bless him. We sure missed him. But we forged ahead and did the best we could.

Interview #1

I mentioned to you . . . a visit the astronauts made to the [LBJ] Ranch in April of '62 . . . I asked President Johnson about that and all he said was, "They came over to air their grievances." Do you recall what the grievances were?

As I recall that trip to the ranch, it kind of got started . . . I don't know whether you remember that period. That was a period when everybody was trying to give things to the astronauts. A fellow down in Houston by the name of Sharp . . . wanted to give each astronaut a house. This was pretty confusing to those guys, because there's nothing wrong with having people give you something except what other people think of it . . . To a lot of people in this country, a house represents almost their entire ability to save all their lives. When they're old, they can finally own their own house. When somebody turns around and gives one away, why, they feel resentful about it. Well, some of the astronauts thought

Dr. Gilruth receiving congratulations from NASA's Associate Director Walter C. Williams, Astronaut Alan B. Shepard, Jr. and Astronaut John H. Glenn, Jr. after President Kennedy presents the "President's Award for Distinguished Federal Civilian Service" during a special ceremony conducted for Gilruth on the lawn of the White House on August 7, 1962. (NASA Photo S-62-04848.)

they ought to be allowed to accept those houses and it was quite controversial. And this was right around the time when, you remember, Deke Slayton was not allowed to fly because some of the doctors in the Air Force, who had gotten on the inside of our selection business, decided that they ought to serve their country by pulling the string on Deke, although they initially gave us assurance that he shouldn't be the first man to fly [in space], but they saw nothing wrong with his flying. And, of course, that's the right answer. There's no reason why he should've been prevented from flying. I fought for him all the time, but I couldn't fight against the M.D.'s when they got together and said it's not the thing to do. So, there was that problem, and at the Gridiron dinner one year when Kennedy was President—I was there and Al Shepard and I think most all the other astronauts, I know Deke was there, too—the President was brought into this. I guess they were a little upset, and the President was interested . . .

But it wasn't publicly discussed.

Oh, no. No, certainly not.

**Oh, I see. That's where the President learned about their griev-
ances, in other words.**

Well, this was one of the places. The way I remember it . . . the
President had asked him [Lyndon B. Johnson] to invite the astronauts to come out to the ranch and kind of let down their hair and
talk about things. And he asked me if I'd mind coming along with
them. So I told him, no, I wouldn't mind coming along. We were
there over my silver wedding anniversary, but I wasn't in a position
not to go. So I went out there with them and we spent a couple of
nights and a day.

That was a long session, then, wasn't it?

Well, we didn't spend all the time talking. We went around and
looked at all the Johnson City things and the various parts of his
ranch and his cattle and picked bluebonnets . . . Well, we did
manage to have a couple of pretty good sessions about this stuff.
And I think it was a good thing to do. This was right at the time
that I told Wally Schirra he was going to fly the flight that followed
Carpenter's flight. No, that Deke would have had the Carpenter
flight. That's right. Deke would have had the second flight. Might
have been Deke's.

Well, he had been assigned, hadn't he? Hadn't he been assigned to . . .

Yes, I think he'd been assigned. And then this thing came up and we
put Carpenter in his place. And then Wally wasn't sure where he
stood . . . So that's how it all happened. And these various points were
covered and Mr. Johnson was very helpful, in helping those boys
come to a concurrence that they shouldn't participate in some of this
stuff. But I think most of them made up their minds . . . But it's nice
to have top management take that much interest . . .

There were a lot of other things, too, that I thought were not as questionable. Some of them invested their money in things like motels. And there's nothing wrong with that as long as they don't use their names and the fact that they own the motel. They ought to be able to invest their money in legitimate ways, the way other people do. You know, those guys have some rights, too. I think some of them went in with motels. I think the rules were clarified under which they ought to operate those things. So, it wasn't all the Sharp's thing, but there were a number of other things at that time that we talked about.

I can remember as late as '67, down at the Cape, if the word got around that an astronaut was in one of the bars, the bar would fill up almost immediately. Word of mouth would pass around, you know, they were still in there. And it was even truer, of course, in '62 than it was in '67.

Oh yes, that was wild down there.

. . .Nobody foresaw how instant fame would accrue to the astronauts the way it did.

Shorty Powers predicted it. I think that kind of guy would know the public mind pretty well, and he predicted it. I remember when he first told the astronauts what their lives were going to be like: that after the flight, after they've been debriefed, the first thing was they'd go to see the President. (And that was pretty damn true.) And that they'd probably appear before Congress and there'd be parades. That's the kind of ballgame we were in back in those early days of space. Of course, we saw it happen in the Soviet Union before we saw it happen in America. I remember seeing the pictures of Gagarin walking along that red carpet that extended about two miles from the airplane to the platform where all the party leaders were. And so they really showed us what the thing was going to be like before we started.

But it was pretty wild in those days when Al first flew and when he went to the Capitol. Webb and I went with him. I'll never

Dr. Gilruth with his mother and father, Mr. and Mrs. Henry A. Gilruth (left) and his wife, Jean, are shown on the White House lawn shortly after receiving the "President's Award for Distinguished Federal Civilian Service" from President Kennedy on August 7, 1962. (NASA Photo S-62-04851.)

forget some of those experiences—going up there with John Glenn. And going to New York. I went to New York with John. Webb said, "Look, you're going to go right along with him and you're going to get every medal that he gets." And this was old Jim Webb, you know. He had in his mind that the astronaut thing was great, but the technical side was great, too. Well, he forgot one thing: in the eyes of the people, the thing that really counts is the guy who puts his life on the line. And they aren't so far wrong on that either. But Jim Webb did a lot to try to see that the astronauts weren't the only people in the whole show. I'll never forget some of those experiences. Some of them I had with Jim Webb himself. I saw him get the Medal of Honor that the City of New York gave him . . . It did my heart good because nobody could have done that lunar project without him.

I remember in the last two or three months of the Johnson Administration, Webb was getting a medal every time he went to the White House. I said to him, "What is he trying to do, make a medal-of-the-month for you?" Well, you know, he got the civilian meritorious medal, or whatever the highest medal the President can give to a civilian.

Medal of Freedom.

Medal of Freedom. He really had tears over that one, because he hadn't expected it. Speaking of fame that accrued to the astronauts and such, what are your views on this stamp deal, the Apollo 15 mess that they got into, which opened up a Pandora's box of a lot of other little items from the past . . . Tell me about how these boys got into such a mess.

I don't know. I really couldn't tell you. I remember telling the crews when we brought them on board—I don't remember which set it was, it wasn't the original ones, it was the ones downstream of that, after I'd some experience—that if there's anything in your past that won't stand the light of day, right now you ought to resign, because it'll come up. There's no way it won't come up, because that's the kind of a ballgame you're in, you're a public figure, just like you're running for president. Furthermore, anything you do in the future, you ought to consider that you do it right on the street in front of all the people in the world in broad daylight. Because if you do anything that isn't completely right, it'll be found out and you'll be in trouble. Some of them might have listened to me, and some of them might have forgotten that. But that's the kind of way they have to live. Unfortunately, most people, if they do some little thing that's wrong, often they're caught and reprimanded without having it come to the attention of everybody in the world. Unfortunately, in the case of the boys in Apollo 15, they did something that, although I wouldn't say it was a crime, it wasn't exactly the right thing to do. And they didn't have any warning, it just went right straight out, and they were judged by it. Everybody in the world heard about it. So, I just feel that it's very unfortunate.

90

Dr. Robert R. Gilruth, director of the Manned Spacecraft Center greets Apollo 15 Mission Commander David R. Scott as he and astronauts Alfred M. Worden (Command Module Pilot), center and James B. Irwin (lunar module pilot) leave the recovery helicopter aboard the *USS Okinawa* after a successful splashdown on August 7, 1971. (NASA Photo 71H-1246.)

It certainly was a lack of good judgment that did it. And I feel very sorry for these boys, because it's awful tough on them, it's probably far more severe than they deserve for what they actually did that was wrong. But you can't blame it on anything other than just being naive, and probably talked themselves into the fact that, well, we're really doing a great thing, and we ought to get a little something for it. You know how you can talk yourself into something like that sometimes. I don't know, I've never discussed this with them, but this is what I think.

You never talked to the Apollo 15 crew afterwards?

I haven't had the guts to talk to them about it. Enough people have talked to them. The Senate has talked to them, and there are enough people that talk to them, and I just feel too sick about it.

Well, of course, you didn't have to because you were no longer the director when this all broke.

I was no longer in the loop, in that sense.

I think I mentioned the other day that the only time that you, Deke, and others had been overruled on assignments of crews, was on Apollo 13 when Alan Shepard was taken off, I believe. Later he was put onto 14.

I guess you're right.

And Deke was the one who told me that George Mueller was the one who forced the change. How did he do that? Did he just telephone you in Houston and say you have to take Shepard off this?

I think it happened in Houston. I'm not real sure about that right now, but I remember Mueller coming and changing some of that in a way that we really couldn't argue against.

I believe his argument was that he hadn't had enough reindoctrination to fly that early.

Yes, that could well be. And it could have been that he was right, because we didn't put up any fight against it. But I think that probably was after 12. What kind of spacing were we flying at in those days? Was it six months?

Yes, 12 was in November of '69 and 13 was in April of '70. So that was just about six months.

I don't know whether we'd already announced crews.

No, you hadn't announced yet.

It was probably at the time we were debriefing the 12 crew there at the lunar receiving lab.

92

Dr. Robert Gilruth (left), director of the Manned Spaceflight Center, and Dr. George Mueller, NASA associate administrator of manned space flight, talk by phone with NASA Administrator James E. Webb and Dr. Robert C. Seamans, Jr., NASA associate administrator at NASA Headquarters. They are discussing the successful conclusion of the eight-day Gemini V space flight following the recovery of Astronauts L. Gordon Cooper Jr. and Charles "Pete" Conrad Jr. from the Atlantic splashdown area on August 29, 1965. (NASA Photo S-65-44213.)

It must have been, because Mueller retired shortly thereafter. I think his retirement was announced within a month after 12 flew.

That was '69.

Yes, so it would have been sometime during that month's interval.

Unless it occurred before that. It could have occurred before 12.

Oh, of course. Thirteen would have been announced before then.

It had to be when we were debriefing after 11, because we had to have more lead-time. I remember being there, of course, with Deke, and he kind of opened the whole gamut up. And our relations weren't awfully good with Mueller at that point. That was around the same time that we took on all the MOL [Manned Orbiting Laboratory] astronauts. And we knew we didn't have flights for all those people, and we told Mueller that, but he still wanted to take them on . . .

I was convinced that Al Shepard would make a very good commander for our mission. I'm very happy to say that he did an excellent job. I'm very proud that the first guy we had in space also was the guy that was able to lead an expedition to the Moon. It was great.

. . . Do you remember on September 17, 1962, you made a speech at the National Press Club[3] in which you said that you trembled at the thought of the integration problems involved in joint missions with the Russians? And three days later President Kennedy came out with his proposal at the United Nations[4] to . . .

Yes, I remember that. And I didn't know what the President was planning . . .

No, nor did anybody else.

. . . and neither did [Hugh] Dryden, who was right there with me at the time, and he sort of agreed with me. Well, I just said what I thought. I still think I was right. We were having enough problems trying to integrate Marshall and MSC at that time.

Well, now, you've been to Moscow on these recent deals. Does it look pretty good to you for integration of the whole thing?[5]

Yes.

Of course, the atmosphere is completely different from what it was in 1962 and 1963.

Well, actually, I led the first two groups over there. I went over there in October 1970,[6] and I guess that's the first time I really thought that

Discussing the scrub of the Gemini VI space flight are (from left), Christopher C. Kraft Jr., Red Team flight director; Dr. Robert R. Gilruth, center director; and George M. Low, deputy director. The three officials monitored the countdown on NASA's Gemini VI from their positions in the MSC Mission Control Center on December 12, 1965. The Gemini VI flight was subsequently rescheduled and launched on December 15, 1965. (NASA Photo S-65-62061.)

we might do it, because of the way they responded. I was told that, gee, that system is an impossible thing. We went over there and I didn't find it was impossible at all. I found the people to be very bright, and that the problems of talking back and forth through interpreters wasn't as tough as I thought it would be. They had an idea of how it could be done and they gave as much as we did. I met Keldysh[7] and he was very interested in it. And he took us all out to dinner. I spent a couple of hours with him in his office talking about all this. And he spoke in English. I was convinced they . . .

His English was good enough not to use an interpreter?

Yes. I was convinced that they were really interested in a project with us and we were able to agree on enough things so we could sign an agreement while we were there, and we were there less than a week. That's hard to do even with an English-speaking group. So I came back and I was very impressed with the situation. We went over again, or they came over here first, as I recall,[*] and then I went back and took a larger group a year later. We were there over a week, and we had working groups and a whole bunch of things. Again, we were very successful at agreeing on things. There were some things we couldn't agree on, but we agreed that we couldn't agree and we agreed to work on those points. They generally needed more lead time than we do because they've got to go back to a bureau. They can't make any tentative agreements. They've got a little different problem than we have. I thought it [the Apollo-Soyuz Test Projects] was going to be a very good exercise for the two countries. I think we've got a lot to gain from it. I think it's going to work. I hadn't thought until this minute that I'd made those remarks about not being able to work with . . . But I sure did. I remember somebody got some poison pen letters of what's the matter with me, disagreeing with the President. Well, I didn't know that he was going to disagree with me.

You know who else didn't know? Webb didn't know, until the day before he made the speech, the 19th of September '63. He was making a speech in St. Louis, and in his prepared speech he had some sort of anti-Russian remarks. Well, he used them because they sold Congress. And he always used the Russians. But he changed his speech before he delivered it because he got a call from Washington saying the President's going to say this. So he was caught in it, too.

Interview #4

Why did you decide to leave MSC?

My wife [Jean] was ill, and I wanted to spend more time with her. I wanted to keep with NASA, but I wanted a job that was less demanding and where I could do what I felt I needed to do. She was ill and we wanted to have time to do a little traveling, visit people that we hadn't seen in a long time. The space program had been very demanding, and so that was the reason. I took this job because I thought I could do some good for NASA. At the same time, I didn't have to work every day. I had lots and lots of leave.

How long were you in that position [director of key personnel development]?

About a year, I guess. But I don't think I had that same title. I think I was a consultant to the administrator or something like that, for the others. I had this one for about a year.

Did you move directly back to [Houston]?

No. I resigned completely before we moved back here.

So you remained down in Houston.

Yes.

So both you and von Braun ended your careers at Headquarters.

But not being there. We reported to Headquarters, but did not live there.

I'd just like to ask you one last question, which is most open and general. As you look back on your career, which is quite a long and complex one, what do you see as the most satisfying part?

I think the most satisfying thing to me is the memory of all of the years and the developments over those years, to have been an active participant in so many of the great things that the United States of America has done in aviation and in space flight. I can't pinpoint any one place that stands out . . . It's fun to look back, and it would be fun to have a more vivid memory of some of it. But I have enough memory to appreciate the opportunities that I've had and the fun that I've had participating in these things.

ENDNOTES

1. The Glennan-Webb-Seamans (GWS) Project for Research in Space History is a series of oral histories conducted by historians at the Smithsonian Institute's National Air and Space Museum. The role of management in NASA is explored through 193 hours of interviews with 22 individuals. These interviews examine various aspects of NASA's management practices in the Apollo program, taking a vertical slice through the organization from NASA's top management in Headquarters to management in the NASA Field Centers. A strategy of the GWS Project was to intensively interview (over 10 hours) key individuals to fully explore the ways in which management practices integrated scientific, technical, and political concerns. Complete copies of the transcripts from the GWS Project can be found on-line at:
   <http://www.nasm.edu/naom/DSH/ohp-introduction.html#GWS>.

2. Democratic Representative Albert Thomas of Houston was chairman of the House of Representatives Independent Offices Subcommittee of the Appropriations Committee.

3. The event referred to here actually occurred on September 17, 1963. Gilruth spoke at a luncheon meeting of the National Rocket Club in Washington. During his luncheon speech, Gilruth reacts to calls emanating from Congress for a joint lunar expedition by saying, "I tremble at the thought of the integration problems of a Soviet rocket with a U.S. spacecraft. . . when I think of the problems we have experienced with American contractors who all speak the same language." He goes on to say, "the proposal would be interesting and significant—but hard to do in a practical sort of way. . . but I'm speaking only as an engineer, not an international politician." David S.F. Portree, *Thirty Years Together: A Chronology of U.S.-Soviet Space Cooperation* (NASA Contractor Report 185707, NASA Johnson Space Center, February 1993), pp. 4.

4. On September 20, 1963, President Kennedy proposes a "joint expedition to the Moon" before the U.N. General Assembly. Ibid., p. 5

5. Soviet and U.S. space officials, led by Gilruth, met in Moscow October 26-28, 1970, to discuss joint space projects, including a common docking system. Three working groups

were established to further study the docking of a U.S. and Soviet spacecraft. David S.F. Portree, *Thirty Years Together: A Chronology of U.S.-Soviet Space Cooperation* (NASA Contractor Report 185707, NASA Johnson Space Center, February 1993), p. 12.

6. Academician Mstislav V. Keldysh, physicist and president of the Soviet Academy of Sciences.

Dr. George E. Mueller, left, associate administrator for Manned Space Flight, expresses satisfaction after the successful July 16, 1969, Apollo 11 launch to Rocco A. Petrone, Kennedy Space Center Director of Launch Operations. (NASA Photo 108-KSC-69P-638.)

CHAPTER 5

GEORGE E. MUELLER

(1 9 1 8 –   )

During his six years of service at NASA, George Edwin Mueller introduced a remarkable series of management changes within the fledging agency during a time when strong leadership and direction were critical to its survival. Mueller joined NASA in September 1963 as deputy associate administrator for manned space flight, a position which soon changed its title in November of that year to associate administrator for manned space flight, a position he held until leaving the agency in 1969.

As associate administrator for space flight, Mueller worked with both the Gemini and Apollo programs and the three NASA centers devoted to manned space flight–Marshall Space Flight Center, Manned Spacecraft Center (now JSC), and Kennedy Space Center. Mueller worked to further the development of human space flight and to improve efficiency within NASA itself. Among his many accomplishments while at NASA was his desire for and receipt of an earlier flight schedule for the Gemini program. In addition, he successfully modified the testing of the Saturn V and the Apollo spacecraft from the previous comprehensive approach to the controversial "all-up" format. Mueller also requested that contractors send completed systems delivered to Cape Kennedy in order to minimize the need for rebuilding. Mueller implemented a concurrency program, which allowed a continued program development if something were to

NASA's Deputy Administrator Dr. Hugh L. Dryden (right) is shown swearing in Dr. George E. Mueller as deputy associate administrator for Manned Space Flight at NASA Headquarters on September 1, 1963. (NASA Photo No. S63-17539.)

happen that seriously jeopardized the schedule. Also known as alternate paths, this program proved invaluable in the wake of the Apollo 204 fire. While workers were investigating the fire and making changes to the Apollo Command and Service Module, others were continuing development of the launch vehicle and the remaining spacecraft such as the lunar module. These changes saved time and money for the Apollo Program, which ultimately lead toward achieving President Kennedy's goal of landing a man on the Moon before the end of the decade.

After receiving his MS degree in electrical engineering from Purdue in 1940, Mueller began work at Bell Telephone Laboratories in New York City where he pursued research in video amplifiers and television links. After the U.S. became involved in WW II, Mueller began work on airborne radars and was soon transferred to Bell's laboratory at Holmdel, New Jersey, where he stayed for the next five years. Mueller left Bell in 1946 and became an assistant professor of electrical engineering at Ohio State University where he earned a Ph.D. in Physics in 1951.

In 1955, Mueller began consulting with Ramo-Wooldridge (the predecessor of TRW) in the area of radar. At this time, Ramo-Wooldridge built missile guidance radars and Mueller was called upon to review their designs and troubleshoot potential problems. It was during this time that Mueller earned the reputation as a problem-solver and developed his theories regarding management that would serve him so well later in his career. Mueller held various positions during his five-year stay at Ramo-Wooldridge, which later became Space Technology Laboratories (STL).[1] It was during his stay at STL that Mueller made his first contact with NASA while working on the design and construction of a series of lunar probes called Pioneer. While at STL, Mueller's achievements include the design, development, and testing of the basic systems and components of the Atlas, Titan, Minuteman, and Thor ballistic missile programs; development of other space-related projects such as Explorer VI and Pioneer V; and establishment of the U.S. Air Force tracking network for deep space probes.[2]

Mueller left NASA on December 10, 1969, to begin work as vice president of General Dynamics Corporation. In 1971, he joined System Development Corporation, a computer software company, where he served as chairman of the board and president until 1980 when he became chairman and CEO. He left in 1983 to become president of Jojoba Propagation Laboratories. Mueller began working for Kistler Aerospace in April 1995 where he remains director and CEO leading this privately held corporation's efforts toward developing a fully reusable multistage launch vehicle.

Editor's Note: The following are edited excerpts from three interviews conducted with Dr. George E. Mueller. Interview #1 was conducted by Robert Sherrod on April 21, 1971, while Dr. Mueller was vice president of General Dynamics. Interview #2 was conducted by Sherrod on March 20, 1973, while Dr. Mueller was president of System Development Corporation. Interview #3 was conducted on August 27, 1998, by Summer Chick Bergen and assisted by Carol Butler of the Johnson Space Center Oral History Project.

Interview #2

. . . I'd like to get it straight on how many missions you really wanted to fly. Of course, you started out going on up through Apollo 20 before the last three were cut. I guess they were cut out after you had departed.

No. the decision was made to cut those out before I left . . . The real problem was the decision that was made to not continue producing the Saturn V, which was made first. Then the question was, since you weren't going to fly a great number of missions, how many should you fly? It turns out that the Houston people were concerned about the safety of flying out to the Moon and back and Bob Gilruth in particular wanted to limit the number of flights to the Moon. But I think that the real forcing function was the scientific community's desire to stretch the time interval between flights and you found that you were going to start flying equipment that was old and the cost of maintaining that equipment over long periods of time was quite clearly going to be high and so we tried to develop a program that would provide a continuing manned space program and bring on new equipment at as early a time as possible. So when we made up the integrated space program we took cognizance of the desires of the scientific community and we tried to arrange a program where there would be no significant gaps in manned space flight. So the plan that Vice President Agnew finally presented when President Nixon first came into office was a plan that we had developed in manned space flight which, among other things in order to conserve funds, we decided that Apollo 18³ would be the last flight. Now I think that was a mistake because I was arguing for a greater frequency of flights and therefore using more of the available hardware. The actual cost of flying these things was really very little. On the other hand, we also needed Saturn V's for the Skylab program and we needed to have at least two of them so that we had a back-up in the event the first one failed.

. . . As a general philosophy, you wanted to fly as few as possible before the landing on the Moon.

Exactly and for a very simple reason—it got you to the Moon quicker.

You wanted to make this deadline before the end of 1969.

Yes. And I particularly wanted to make sure that we kept the enthusiasm of the people up as we were going and to reduce the risks to an absolute minimum of failure before we got to the Moon and back. The more times you fly out there the more probability there is you won't come back.

I remember that you argued against the F mission—Apollo 10.

Yes.

Do you still think that it was unnecessary to fly it?

I think that we learned a fair amount on that flight. I suspect that we could not have done without it but on the other hand it was a worthwhile effort, particularly since the lunar module was far enough behind so it couldn't land. We needed to maintain the momentum of the program.

What about the C-Prime mission [Apollo 8], which they pulled on you while you were in Vienna. You were not very much in favor of this mission were you?

As a matter of fact, I was quite in favor of it. I thought it was a great idea. But, I used it as the mechanism for forcing, however, a very

Dr. George E. Mueller, associate administrator for Manned Space Flight, NASA, follows the progress of the Apollo 11 mission. This photo was taken in the Launch Control Center at the Kennedy Space Center the morning of the launch. (NASA Photo 108-KSC-69P-647.)

careful review of the entire program. After all, it was the first time since the fire [Apollo 1] that I had people who wanted to do something who were pulling for progress. Prior to that time I had to provide all of the push . . . So I was able to take advantage of that and really force a very careful review of everything that went into the success of that mission. I suspect that without such a thorough review, we might well have had problems downstream; you know problems that aren't immediately apparent tend to come back and bite you later. We made sure that we looked at everything that we could possibly think of that could go wrong and then made ourselves positive that if it did go wrong, we still wouldn't lose a crew.

Yes. However, the whole Apollo program would have gone down the drain if an oxygen tank had gone up on that mission and you didn't have the LM for a lifeboat.

Yes, although actually the most difficult mission in my view was still 501 [Apollo 4—the first launch of the Saturn V]. The whole Apollo program and my reputation would have gone down the drain if there had been an oxygen tank blow-up on 501. The whole concept of all-up testing, which made it possible to carry out the program, was in grave doubt up until 501.

Yes, that's the one that had to work.

Yes.

Although you wouldn't have lost any men on it if it did fail.

True enough, but you probably would have had a great deal of difficulty getting any man on one of those vehicles, too. Interestingly enough, 502 [Apollo 6] had everything in the world go wrong with it and if that had been 501, we probably would have had a great deal of difficulty proceeding . . . That also incidentally, was one of the reasons that I was so anxious to have a chance for a very thorough in-depth review of that program, which Apollo 8 provided us with.

To be retrospective, what about the scientists' argument? You remember in the summer of '69 particularly going into 1970, the scientists were making speeches, giving interviews, and writing articles about how there is no scientific aspect to the Apollo program.

It's very interesting. About a year after the first landing, and after I was out of the program, one of the chief critics of the scientific content of the program came around and said to me, "You know, we sure made a mistake in not following your advice. We should have had these vehicles flying on three month centers. We would have learned so much more if we had more time on the Moon." Their argument was, "Gee, you need time to digest what you've learned." The fact of the matter is there is such a variety in the areas of the Moon that they weren't able to understand what they got from one area without having information from the other areas as well. So it wasn't until fairly near Apollo 17 that there was even a beginning of understanding of the lunar geology. Well, now that's an overstatement. Yet in a very real sense I believe that we would have had a much better understanding of the Moon, we would have ended up with a great deal more information if we had been able to fly those missions on three month centers and fly more of the vehicles.

Interview #1

. . . I understand that in the all-up testing, you were the one who really put it over in NASA. It was not yours originally, though. The Air Force had used it before, had it not?

We had introduced the concept at STL [Space Technology Laboratories] of all-up testing and I was involved in the development of that concept. It had been introduced in the Air Force although it was just about that time that it was first being tried . . . I had also introduced the concept of alternate paths so that in the event something went wrong with the Saturn V we could carry on the program in the Saturn I, or if something went wrong with the spacecraft, we could nevertheless carry on a limited program in developing the Saturn V. We had a number of alternates. We had a strategic plan of logical decision points and we used it.

In what case did you use it?

108

For example, on the occasion of the fire, which slowed down the [development of] the spacecraft, we were able to continue the development of the launch vehicle and fly some early spacecraft in the unmanned mode which permitted us to move forward even though we were still making changes in the spacecraft to accommodate the fire problem. And we dropped out a number of things in the Saturn I program as a consequence. We freed up a lot of hardware because the Saturn V was successful. If the Saturn V had not been successful, we nevertheless could have developed a spacecraft on Saturn I's completely.

You said that after the fire you continued launch vehicle development. That means Apollo 4, 5, and 6. They weren't changed because of the fire were they?

They were rescheduled. We lost almost a year in that process. Because of that year we dropped out a series of unmanned flights on the Saturn I.

Who derived the strategy of making the Phillips Report a set of notes in the congressional hearings?[4]

I think it was either Jim Webb or me.

Jim says he didn't know what they were talking about until February 27, 1967.

I know he didn't know what they were talking about. Neither did I. I'll agree because I didn't think those notes or the report existed, so I was astonished when I got a copy of it.

You thought the notes had been written up, or the report had been written, but simply hadn't gone beyond one or two copies?

I thought they had been destroyed, again because they didn't appear to be constructive in terms of solving the North American problem. But we did pull out the constructive part, to give a series of brief-

109

ings. And we knew about that—I knew about that because I said, "I don't agree." I guess in all honesty I never saw the notes, read the notes, or read the report (except to look at one of two sections) . . . I told Sam [Phillips] that I didn't think they were appropriate then to publish and get into circulation. And I think that turned out to be a correct assessment. But on the other hand, we did have a good set of notes, which had been used to brief the people at North American and the people at NASA Headquarters. I guess in all honesty, I wasn't sure from the Mondale question, which is where this whole thing started, just what in the world he was talking about.[5] I thought he was talking about something else. I couldn't quite relate—because he had the wrong date, among other things.

Yes, I think he said the summer of '65 instead of November or December. Mondale brought it up on February 27th and it lay dormant until April 11th when it came up in the Senate again.[6] But on February 13th, just two weeks before the 27th, Jules Bergman[7] saw a copy of it over at the Office of Manned Space Flight [at NASA Headquarters] and told Mondale about it...

Jules saw it in the Office of Manned Space Flight?

So he told me.

Well I'll be darned. I assumed he'd seen it out at North American . . . I'm surprised at that because I didn't know it existed at that time.

Somebody told him about it, and he called it the "Sam Phillips Report," which is what Mondale called it when it first came up.

You know, it may be interesting and fascinating like any other thing, but in terms of what we learned from it [the Apollo 1 fire] in future programs—we probably spent more time on it per unit information gained for doing better in the future by several orders of magnitude than on a whole host of other things that really made it possible to do the program.

110

Apollo 11 mission officials relax in the Launch Control Center at the Kennedy Space Center following the successful Apollo 11 liftoff. From left to right are: Charles W. Mathews, Dr. Wernher von Braun, Dr. George E. Mueller and General Samuel C. Phillips. (NASA Photo 108-KSC-69P641.)

Well, it caused a general tightening up throughout all of NASA and all of the contractors.

Yes, it took about two years to get it to the point where we could fly again with any reasonableness. And if you'll recall, I had to take the lead in convincing people that it was safe to fly and that we really couldn't afford not to take some risks, that there wasn't any way to fly any of these things without risk. And there was a whole year where nobody was willing to take any risk whatsoever.

Speaking of taking risks, I've seen this point made in only one place. Webb made it during his congressional testimony. He said, "I wonder now why we ever planned to fly the Block I spacecraft at all." Why was it—all of them were scrubbed except this one (Spacecraft 012—Apollo 1)?

Well, it's a good question. In fact if you go back, you'll find that we almost scrubbed 012, and the reason we almost scrubbed it was that it wasn't clear that we were going to gain enough in flying it to make it worthwhile finishing it up. I was on the side saying "Well, I don't think the forward program will be helped enough by the extra effort required to build this—get it built and furnished in a way that it could fly—but what we ought to do is scrub it and do the next one right." However, Bob Gilruth and Joe Shea and the Houston people said, "No, we'll learn an awful lot by moving this thing on through, even though it isn't exactly the same configuration it clearly tests all of the equipment and all of the launch apparatus and so on, so we'll get that out of the way—that in the long run, will save us time rather than lose us time." And I went along with them because it was a valid argument. We had an agreement, however, that if it was going to slip any more, more than another couple of months, that we would in fact bypass it. None of us, of course, had any idea that this would pose any danger to the crew. We all thought that the thing was perfectly safe. And in fact it had been through design certification reviews which supposedly picked up these things. And one of the things that was asked was "Has this thing been examined for being fireproof?" You'll find in the record that there was a report prepared which said yes it met all of the needs for fire resistance.

I've heard it said about the 204 fire hearings that you and Webb disagreed on how to testify. Your idea was to get the hearings over with and get on with business, and Webb knew or felt that NASA had to wear this "hair shirt" with a "mea culpa" attitude. Is this a fair statement?

I think it's a fair statement. Jim, who's had much more experience in this area than I did, felt that it would take a number of months for all of the facts to filter up to the surface. And if we finished the hearings too soon and then further facts were developed, you'd be terribly vulnerable to being accused of whitewashing a terrible situation. He wanted the hearings to last long enough so that every single fact that could be deduced was in fact brought to the surface.

He trotted out so everybody could see it and understand it and everything hidden turn up. That strategy surely worked. It was helpful, because it took two months for some of the things that were lurking in people's minds to come out. I guess the hearings themselves lasted four months, and that seemed a bit long-winded. In one other area, I guess with respect to the Phillips Report, we did disagree. The minute I discovered the Report, I told Jim about it and we were both concerned and disturbed because in good faith, we told the Senate committee that we didn't know of any Phillips Report, and then here is this report on North American Aviation signed by Sam Phillips. And that was, I guess, two weeks after Mondale's question [during the February 27 hearing] when this thing came to light. That's why you astonished me when you said Jules Bergman said he saw it at the Manned Space Flight Headquarters.

How did you and Webb disagree?

I felt that we needed to protect the contractor from a full disclosure of the material in that report. Not because it was North American but because of the whole process. This kind of management review depends upon finding out what the truth is and you can only find that from the contractor and he had given it to them in confidence so we ought to keep it in confidence. So that was my initial reaction, and Sam Phillips felt very strongly about it . . . But at that time we didn't know, had no idea, how much was known about this. About two or three days later I had a chance to talk with Jim Gehrig[8] (Jim Gehrig asked me and I told him about these notes). So then we had a long discussion about how it would be better to summarize it, give a summary to the committees so they would have it available. We made the whole report available to Tommy Thompson [Thomas H. Thompson, director, Apollo Systems Engineering] as soon as we discovered it, so they had it for their review.

Of course you know that when Gehrig first found out about it he asked the General Accounting Office (GAO) for it—that's how it got bound up and labeled with "Report" on the front at that time.

That was later . . .

No. The GAO got it before Mondale ever brought it up. That surprised me when I learned it.

Really?

Yes.

I didn't know that.

I think it was about February 17.

Before Mondale raised the question?

Before Mondale raised the question. I think Gehrig himself asked the GAO to look into it.

You may be referring to the second time that Mondale asked the question.

No. I'm talking about the first time. Before February 27, 1967, when it first came up in the Senate.

The GAO then knew more than Jim Webb or I did.

Interview #3

. . . Why did you decide to leave NASA and go back to industry?

Well, several reasons. One is that the decision had been made to terminate the Apollo program, and that was a good time then to leave before, and let someone else take over for the next phase. From a practical point of view, I needed to go make some money so I could keep my family going. It was costly for us to join the Apollo program. My salary was half what I was making in industry when I went there, and it was just a strain to keep the family going and work going at the same time. So I went back to industry.

It was nice to leave with the triumph of landing on the Moon.

Well, it was a good time to leave in that sense. You know, it looked like it would be another 5 or 10 years before the next program was going to come to fruition. There's also the general thing that if you stay in Washington long enough, if you do anything, you create enough enemies to make it difficult to get anything done. I'd left before I think I created that set of enemies, but it's clear that you have a limited time of effectiveness in Washington if you really are doing anything. If you're not doing anything, you can stay there indefinitely.

ENDNOTES

1. STL [Space Technology Laboratories] was originally a spinoff of Ramo-Wooldridge in 1959 and later became part of Thompson-Ramo-Woolridge (TRW). STL was a corporate think tank that supported the Air Force Ballistic Missile Program. In 1960, a group from STL broke away and formed the Aerospace Corporation. In 1966, STL was renamed TRW Systems group. Courtney G. Brooks, James M. Grimwood, and Loyd S. Swenson, Jr., *Chariots for Apollo: A History of Manned Lunar Spacecraft* (NASA SP-4205, 1979), p. 200.

2. Henry C. Dethloff, *Suddenly, Tomorrow Came: A History of the Johnson Space Center*, The NASA History Series (Houston: NASA, Lyndon B. Johnson Space Center, 1993), 101; George Mueller, interview by Howard E. McCurdy, June 22, 1988, transcript, NASA Historical Reference Collection, NASA Headquarters History Office, Washington, D.C.; Interview by Martin Collins, July 27, 1987; Henry C. Dethloff, *Suddenly, Tomorrow Came: A History of the Johnson Space Center*, The NASA History Series (Houston: NASA, Lyndon B. Johnson Space Center, 1993), 101; Dr. George Mueller, interview by Martin Collins, February 15, 1988, transcript, NASM Oral History Project, [NASM Homepage], [Online], (September 6, 1996—last updated), available at http://www.nasm.edu/NASMDOCS/DSH/TRANSCPT/MUELLER.HTML [July 2, 1998—accessed].

3. (*Editor's note: Uncertain of Mueller's reference to Apollo 18 as being the last flight and feel that he is actually referring to Apollo 17*) By 1970, NASA began canceling later Apollo flights, specifically, Apollo missions 18-20. On January 4, 1970, NASA Deputy Administrator George M. Low told the press that Apollo 20 had been canceled. On September 2, 1970, NASA Administrator Thomas O. Paine announced at a

Washington news conference that Apollo missions 18 and 19 would be canceled because of congressional cuts in Fiscal Year 1971 NASA appropriations. Remaining missions were then designated Apollo 14 through 17. UPI, "Apollo Missions Extended to '74," *New York Times*, Jan. 5, 1970, p. 10; NASA Administrator Thomas O. Paine in NASA News Release, "NASA Future Plans," press conference transcript, Jan. 13, 1970; "Statement by Dr. Thomas O. Paine," Sept. 2, 1970; *Astronautics and Aeronautics*, 1970 (NASA SP-4015, 1972), pp. 248, 284-85.

4. After the Apollo 1 fire, a review board was established on January 28, 1967, by Jim Webb and NASA Deputy Administrator Robert C. Seamans, Jr. Webb and Seamans asked Floyd L. Thompson, director of NASA's Langley Research Center, to take charge as chairman of the Apollo 204 Review Board. On November 22, 1965, at Mueller's request, Apollo Program Director General Samuel C. Phillips, initiated a review of NASA's contract with North American Aviation to determine why work on the Apollo Command and Service Module (CSM) and the Saturn V second stage (S-II stage) had become so far behind schedule and over budget. On December 15 of that same year, Phillips provided a set of notes which comprised their review to North American Aviation President J. Leland Atwood. These notes, which were highly critical of North American's performance on the Apollo program, later became known as the "Phillips Report" during the Apollo 204 fire investigation. The Phillips Report took on added significance and became highly controversial during the 204 fire investigation hearings when it was revealed that NASA Administrator James Webb was apparently unaware of the existence of the report. John M. Logsdon, editor, *Exploring the Unknown Volume II: External Relationships* (NASA SP-4407, 1996), pp. 527-538.

5. On February 27, 1967, NASA officials testified in an open hearing of the Senate Committee on Aeronautical and Space Sciences on the Apollo 204 fire. It was during this time that Senator Walter F. Mondale, one of the Senate Committee members, first introduced the Phillips Report while raising questions of negligence on the part of NASA management and the prime contractor North American Aviation. It was during this time that Webb and others on the NASA review board expressed doubt and confusion as to which report Mondale was referring.

6. On April 11, the Senate held another hearing before the Committee on Aeronautical and Space Sciences concerning the Apollo 204 fire.

7. Jules Bergman was a reporter for ABC and extensively covered the manned space program for ABC News through to the space shuttle program.

8. James J. Gehrig, Senate staff director.

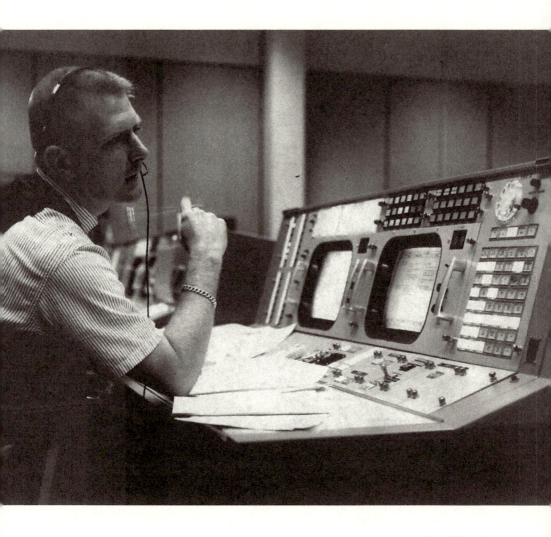

Eugene F. Kranz, flight director, is shown at his console on May 30, 1965, in the Mission Operations Control Room in the Mission Control Center at Houston during a Gemini-Titan IV simulation to prepare for the four-day, 62-orbit flight. (NASA Photo S-65-22203.)

CHAPTER 6

EUGENE F. KRANZ

(1 9 3 3 –)

From a very young age, Eugene F. Kranz developed a unique interest in space flight. Born in Toledo, Ohio, on August 17, 1933, Kranz formerly declared his interest in the subject by writing a high school thesis which explored the possibilities of flying a single-stage rocket to the Moon. However, after graduating from Parks College of St. Louis, Missouri, with a BS in Aeronautical Engineering, Kranz's interests became more down to earth as he shifted from space travel to aviation.

After entering the Air Force and serving in South Korea, Kranz began work for McDonnell Aircraft testing missile launches off of B-52s at Holloman Air Force Base.

Responding to a "help wanted" ad from NASA in Aviation Week, Kranz soon found himself employed in 1960 with the newly formed Space Task Group at NASA's Langley Research Center in Hampton, Virgina. From 1960 to 1964, Kranz worked in the Flight Control Operations Branch developing and writing rules used by flight directors during manned space flight missions.

Throughout his distinguished NASA career, Kranz took on progressively larger and more capable roles within the general arena of spaceflight operations. When the Manned Spacecraft Center opened in Houston (now the Johnson Space Center), Kranz moved to Texas and became chief of the Flight Control Operations Branch. In Houston, he served as Gemini flight director from 1964 to 1968. Between 1969 and 1973, his other duties at MSC include: chief, Flight Control Division; flight director for the Apollo and Skylab programs; flight director for the first lunar landing (Apollo 11); and flight director for the return of the Apollo 13 crew. He also served as the flight operations director during the Skylab program from 1969 to 1974. At the end of Skylab in 1974, Kranz was promoted to deputy director of Flight Operations and then in 1983 to director of Mission Operations. In 1994, he retired from NASA.

Kranz is currently writing a book about his experiences. He lives with his wife Marta in Dickinson, Texas.

Editor's note: The following interview with Eugene F. Kranz was conducted on January 8, 1999, by Rebecca Wright, Carol Butler, and Sasha Tarrant of the Johnson Space Center Oral History Project. The interview begins with Kranz telling of his role at the beginning of the Apollo Program, concluding with a detailed account of what it was like in the Mission Control Center when Apollo 11 landed on the surface of the Moon.

I was working as a flight director on the Gemini IX mission, and it seemed almost overnight I was picking up the responsibilities for the Apollo Program. [Christopher C.] Chris Kraft, myself, and John Hodge were to be the first flight directors to fly the first manned Apollo mission. The remaining flight directors, Glynn [S.] Lunney and [Clifford] Cliff Charlesworth, continued to finish the Gemini Program, and as they continued these missions, I'd go back in and fill in the night shifts for them to give them a bit of a break so they didn't have 7, 8, and 10 days of two-shift operations continuous.

In the meantime, I was a division chief, had the responsibilities for the Flight Control Division, and we were in the process of learning the Apollo spacecraft. We had a new contractor. We were very comfortable working with the McDonnell contractor in Mercury and Gemini. We had developed a very close association with them. They knew us, we knew them, and we believed they understood the nature of the space flight business, where Rockwell had been producing aircraft, fighter aircraft in particular, and were actually getting into the space business, and it ended up in some very heavy, some very strong conflicts of opinions on how the work was to be done.

In particular, with my control team, I demanded the responsibilities to do all of the mission preparation, mission design, the writing of the procedures, the development of the handbooks. This was alien to Rockwell, because they were used to producing all of these products for the people who flew their airplanes. So this was a time period where it just seemed that it was a very tempestuous, very tumultuous time frame, but we managed to get working through this, and it really amazed me how quick the turnaround was between the end of Gemini and the first of the Apollo program. At times I felt like I had a foot in each program and couldn't quite figure out which way to go.

It was also surprising to us that we were going to fly the first of the Apollo series in a manned fashion. Always before, we had been very conservative in the development of our test programs so that we'd do what we call "incremental flight testing." We'd basically take a step at a time and very conservatively make sure that we understood what we had learned from the previous flight. We'd have relatively long gaps between missions—when I say long gaps, I'm talking the order of a few months—and then develop the next mission, which was a small, baby step forward.

But for the Apollo program, the race to the Moon was very real to us that time. We had the lunar challenge laid out in front of us. We were three years from the end of the decade, so this did not allow too much procrastination in the directions that we were taking. So we elected to fly our first spacecraft, I mean first the Block One spacecraft, in a manned fashion with [Virgil] Gus Grissom, [Edward] Ed White, and Roger Chaffee. We had spent time working and training and conducting various tests with back-up crew, and the initial back-up crew was [James A.] Jim McDivitt, and then late in the year 1966, he was swapped out, and Wally Schirra became the back-up crew for the mission.

So we spent a lot of time with the crew really getting to know the spacecraft. The first flight in a program is very intense, and it's the first big milestone that kicks off the program. So there's a lot of personal contact between astronauts, flight directors, flight controllers, and we're all pooling our knowledge, trying to get to the point where we all feel that we're ready. And this feeling just never quite seemed to get there. It always seemed that every time we'd turn a corner there were things that were left undone or answers that we didn't have or we were moving down a wrong path, but we had the confidence that we'd been through this before. We'd been through it in Mercury, we'd been through it in Gemini, so we had the confidence that by the time we got to launch date all the pieces would fit together.

So we continued the testing program, and January 27th, I believe it was, when we were conducting the plugs-out test with Gus Grissom, Ed White, and Roger Chaffee, we had had a test two days before, the plugs-in test, where we had a variety of problems.

We had problems with communications and life support, numerous deviations to our procedures right on down the line, but again we staggered through this test. An 8-hour test would routinely, in those days, take 24 hours, and then you'd take sort of a break because you only had one test team, and you just had to get away from it all. You had to regroup, basically rewrite the procedures, give everybody a little bit of time off, and then come back to the console and hit it again.

The test with Grissom, White, and Chaffee started off early in the morning. I had the responsibility for the Mission Control team, the Mission Control Center, communications, the remote site teams, etc. So basically I was generally the first one in to support the countdown, and we were following—myself, [John] Hodge, and [Chris] Kraft were following the same sequence that we did for the Gemini missions. I would work all of the systems-type issues, John Hodge would do the planning, and Kraft would conduct what we called the execute shift, the dynamic shift with the crew awake.

I had checked out all of the communications. I'd talked with the launch team down at the Cape. We'd picked up all the deviations to the procedures and had worked through into the early afternoon time frame when John Hodge came over. I had handed over the console responsibilities to John, and he was going to continue the countdown until the time frame when we got very close to the simulated launch, and Kraft would come in and pick up the count.

I had handed over to John shortly after noon at this time. The crew had entered the spacecraft, and everything looked like it was going reasonably well for a change. We had had problems in communications, but nowhere as severe as we'd had in the previous day's testing. I went over to the office, and we had small intercom boxes where we could listen to what was going on in Mission Control in our offices, and this was pretty much the norm, and you would listen to what was going on until it was time for you to come over and do your thing or pick up your shift.

By the time that it got to about 3:00, 3:30 in the afternoon, it seemed to be going okay. We had had some problems for holds with the crew, but we were again slowly staggering through the countdown to the point where I was sure we'd finish the test that day.

Marta [Kranz's wife] had our third child, so I'd promised her an evening out for a change, and one of the places that you would go would be the Houston Ship Channel. This is where all of the flight controllers would go. It had a Greek restaurant there, and at that time it was a great thing, apparently, to eat food wrapped up in grape leaves. I'd never done this, but we decided we were going to do this. We were in the process of dressing to go out, and my next-door neighbor, Jim Hannigan, came over. Actually, we were waiting for the babysitter, and we heard some loud pounding on the door.

I went downstairs, half dressed, and I was expecting the babysitter, and instead it was a neighbor. Jim Hannigan came in and identified that he had heard over the radio that there was a serious accident in the launch complex and suspected the crew was dead. So this was the first indication I had that we had the Apollo 1 disaster.

So I came out, hopped in the car, and it's about a 15-minute drive out to Mission Control from the small community that we live in. I arrived out there, and they had secured all of the doors, and there was no way to get in on the phone from the security people up to the Mission Control floor. I kept circling around the building, and there was a freight elevator back there, and I sort of buffaloed the security guard to get access to the freight elevator and up to the floor where we were conducting the test, and got up there and got into Mission Control.

I've never seen a facility or a group of people, a group of men, so shaken in their entire lives. Kraft was there. He was on the phone down by the flight surgeon talking to the people down at the Cape, I believe [Donald K.] "Deke" Slayton. John Hodge and I had grown up in aircraft flight tests, so we were familiar with the fact that people die in the business that we were conducting, so we had maintained maybe a little bit more poise relative to the others, but the majority of the controllers were kids fresh out of college in their early twenties. Everyone had gone through this agony of listening to this crew over the 16 seconds while they—at first we thought they had burned to death, but actually they suffocated, but it was very fresh, very real, and there were many of the controllers who just couldn't seem to cope with this disaster that had occurred . . .

We used to congregate, as flight controllers, over in the Singing Wheel, which was our watering hole. We also called it the Red Barn. One by one, as the controllers secured their consoles, we secured all the records, there wasn't anything more we could do. We went over to the Red Barn, and the proprietor, Lyle, over there also had heard that we had a bad time out in Mission Control, so he sent out all of his normal customers, and for the remainder of the evening we just sat and commiserated with each other and tried to work through the fact that we had lost our first crew. One by one, we all dribbled home.

The next morning, we came back out to work again, trying to see if there were any answers, because in that kind of an environment you're trying to find answers, you're trying to find out why, what happened, etc., and there were no answers. We worked through the Sunday time frame, again just sitting in offices almost just paralyzed. We were so stung.

Come Monday, I had the responsibility for the flight controllers. I was the deputy for the Flight Control Division. I was the chief of the Flight Control Operations Branch, and I wanted to get my people together and just talk to them, because I had also flown fighters over in Korea, and when you lose a crewman or one of your squadron mates, you sit down and talk it over. You've got to get through this. You can't avoid it. The fact is that someone has died. I wanted to call my flight controllers together, and John Hodge decided it would be good to call the entire division together.

So we had a meeting in the auditorium, in Building 30 [Houston] what we used to call the office wing there. John Hodge sort of opened this session. He really reiterated what we had learned about the team, what we had learned about the accident so far, but, in particular, the team that was forming to take charge and determine the cause, determine what we were going to do about it. He talked about the responsibilities of Sam Phillips. He talked about the responsibilities given to [Dr.] George [M.] Low; he was picking up the program. Frank [F.] Borman, [II] was going to be part of a committee, and a guy by the name of Floyd [L. "Tommy"] Thompson was actually assuming the responsibility to lead this group. He came out of, I believe, Langley or—I believe it was Langley at that time.

So this was sort of the news of the day, and it then came time for me to speak. I tend to be maybe one of the more emotional of the controllers. I believed that that's part of a leader's responsibility, to get his people pumped up, and I gave what my controllers came to know as the "tough and competent" speech, and concluded the talk identifying that the problem throughout all of our preparation for Apollo 1 was the fact that we were not tough enough; we were avoiding our responsibilities, we had not assumed the accountability we should have for what was going on during that day's test. We had the opportunity to call it all off, to say, "This isn't right. Let's shut it down," and none of us did. So basically the toughness was from that day forward we would stand for doing everything right, literally being perfect and competent.

We had become very complacent about working in a pure oxygen environment. We all knew this was dangerous. Many of us who flew aircraft knew it was extremely dangerous, but we had sort of stopped learning. We had just really taken it for granted that this was the environment, and since we had flown the Mercury and Gemini program at this 100 percent oxygen environment, everything was okay. And it wasn't. And we had let the crew literally paper the inside of the spacecraft with Velcro . . .

I had each member of the control team on the blackboards in their offices write "tough and competent" at the top of that blackboard, and that could never be erased until we had gotten a man on the Moon. I believe that set the framework for our work in the weeks and months that followed.

It's amazing how NASA took charge of itself in those days. We had pure, raw leadership, incredibly talented and capable people, and by November, ten months later, we were launching our first all-up Saturn V, a very gutsy move by George [E.] Mueller to conduct what was called "the all-up testing," and this was every time you fly you're going to test everything. You're going to test all three stages of the booster. You're going to test the spacecraft. You're going to test inside the spacecraft, the guidance and navigation controls. There is no test that will not be a complete entity.

The obvious advantage of this was, if you're successful, you're buying yourself time on the schedule. If you see a bunch of problems, you've got time to fix them, but if you're unsuccessful, you've got a whole bunch of space hardware that's reduced to junk. So it's a go-for-broke kind of approach that he kicked off that really paid off and, I think, was the real key in getting to the Moon.

Well, his first all-up test was the first launch of the Saturn V in November of the same year that we had the Apollo 1 fire. NASA in those days seemed to have an ability, an incredible ability, to pick itself up and set itself on the right path and then go do things.

The [first Saturn V flight] was Apollo 4, followed by a mission I flew in March. The lunar program, the lunar module, was starting to get behind in manufacturing and tests. The program looked like it was starting to stretch out and the lunar module was going to be the pacing entity. We conducted an unmanned test of the lunar module [flown onboard Apollo 5], which, to me, was probably the most interesting of the unmanned flight tests that we ever flew. Most of the histories don't talk about the unmanned missions, and from a flight control standpoint they're the greatest thing that we've ever done, because if you really think about it, in the test program we are the first people to fly the spacecraft. It isn't the astronauts.

This test again showed the ability of the control team to take charge. We had to launch it. The basic objective was to test all of the propulsion systems in the lunar module, the descent engine that we would use going to the Moon, and this was a throttlable rocket engine. It was one of the first throttlable rockets, liquid rockets, ever developed.

Then we had fire-in-the-hole staging in case we'd have to conduct a lunar abort while that descent engine was still running. We would blow some pyrotechnic bolts, separate the two stages, and ignite the ascent engine, change guidance systems, electrical power systems, and actually launch back off that platform, while we're still going down, launch back up for an abort. This is exactly the same technique that we use when we're on the surface of the Moon. So we had to test that out. And then we had a series of ascent engine burns. This is the engine that we used to get off the Moon.

So this was our baby, and it was a go-for-broke test, again in Mueller's all-up test concept. If we didn't get this done, we weren't going to be able to fly the lunar module. So we launched, and the mission was designed to be only about eight hours long, so we only had one flight control tape. Well, in the middle of the third orbit, when we were getting ready to kick off this testing, we found a computer bug. Somebody had left a fault in the design of the computer software because we had ignited the engine, and as soon as the engine started, it shut down.

This was really key, getting that engine started back up again, and we tried to get it started under automatic control, but in those days you didn't have global communications. You didn't have satellites. You had basically isolated stations, and for an eight- or ten-minute period every orbit one station might be able to track or two stations or three stations. So you had to try to plan your activities all to occur over the stations, and by the time we got to the third orbit we were starting to get into the area of sparse network coverage.

So we took over ground control, just like an astronaut would do in flight. We were punching in the commands for the engine just as the astronaut would do, and we finally managed to get the descent engine testing, all those engine tests accomplished. We then went into the ascent fire-in-the-hole stage, and this is a very complex maneuver. We got that accomplished, but as we finished up, the lunar module had lost almost half of its weight. So we ended up in an autopilot mode that was incompatible. So we started hosing out the fuel, and we were blowing thrusters off the spacecraft, etc., and we had to find some way to again control it, so we shut down one of the propulsion systems to reserve that, came up with a new game plan.

In the meantime, the media was very interested in this, because the east coast newspapers already had their deadlines, and they had declared this mission a total failure, and they had talked about the great setback to the Apollo program. But in the meantime, we kept charging on. We finally got the ascent engines started and completed those tests so that about eight hours after we had this initial problem, at about midnight we finished up with the mission. The west coast newspapers were able to talk about the Apollo program being on track, and that the test of the lunar module was a total success.

Every team had those kinds of experiences. The next mission [Apollo 6] was one that was really a solid test of the Saturn launch vehicle where we lost two engines on one side during powered flight during launch, and the Saturn's not supposed to fly. It's supposed to basically start spinning, lose control, but this Saturn didn't. And when you lose two engines one size, we go through it, we use a term "chi freeze" which means we actually hold the attitude we're in. Well, this kept the booster going essentially straight up.

At the time that we now finish first-stage burn-out, the booster recognized it was too high and too slow, so now it takes all of the five engines in the next stage and starts driving back down towards the ground. Okay. Then the guidance realizes it's got the right velocity but the wrong altitude, so this thing keeps turning around on it.

This rocket that we're getting ready to launch people to the Moon on goes into orbit thrusting backwards. In the meantime, it's throwing all these wifferdils all over the plot boards, and the team, by all rights, they should have called an abort, but since the engines were running, they were just watching this thing keep going.

Well, this is a good demonstration that the Saturn people and the people down at Huntsville [Marshall Space Flight Center] had done a spectacular job of designing and implementing, with huge margins, this rocket. Each of these lessons that we learned in the unmanned missions were very important for us to get confidence in the hardware that we're now ready to put the men on.

We then went through the Apollo 6 and eventually into the Apollo 7 mission, and this was our first manned flight test. Again, myself, Glynn Lunney, and Gerry [Gerald S.] Griffin were the flight directors for this mission. Kraft, by this time was targeted to be a flight director for the early space program, but with the accident and the delay and basically the replanning of the entire program, he actually moved out of Mission Control and just ran the organization that was trying to put all of these pieces together.

I should back up, really, to before the Apollo 7 mission in April of 1968. Kraft, at one of his staff meetings, started talking about the problems that came out of this booster that went into orbit backwards and all the things that had to be fixed, and he started getting concerned about the schedule and our ability to go to the Moon. What he started was a series of mission-planning exercises to determine if we can't fly this sequence that we had planned, what alternatives do we have for maintaining on schedule?

One of the missions that we had in this very long plan sequence was what we called the "E Mission." We gave letter designations to each of the missions. The C Mission was the first command and service module. The D Mission was the first mission involving a lunar module in a manned fashion and the command module, and the E would take this lunar module and the command module into a very high elliptical orbit, about 4,000-mile-high orbit.

Kraft and many of us felt that this was sort of a nonsensical mission. It was just too darned conservative for the tight program that we were flying, and it was really going to delay us getting to the Moon. So he proposed that we take this E mission, which would go the 4,000 miles, only instead of taking it 4,000 miles, take it 250,000 miles up to the Moon. So what we would do we would fly one mission that would go up around the Moon and come back to Earth.

Well, this was very important because it would allow us to check out the ability to use the Saturn to inject towards the Moon, which we had never done. It would allow us to check out the navigation system onboard the command and service module. It would allow us to check out our ability to track from the ground a spacecraft 250,000 miles away.

So he'd kicked off this activity, and Johnny Mayer, who was chief of Mission Planning, was an incredibly gifted individual. Well, he looked at this now as his opportunity to really kick into high gear in planning the lunar mission. All through the months of April, May, June, and into July, we had people within flight control and Johnny Mayer's division seriously looking at a lunar mission at the earliest possible date at the time that we were scheduled to do this E Mission.

Gene Kranz working at his flight director's console in the Mission Operations Control Room at Houston circa 1965. (NASA Photo S-65-60057.)

In the meantime—and this was one of the great things about NASA—you would do many things. You'd be doing one thing with your right hand, another thing with your left hand, and at the same time you'd be dancing a jig. We were in the process of getting ready to fly Wally Schirra's mission, and Schirra would be scheduled for October, I believe.

Come August, I get a call to go over to Kraft's office, and Kraft says, "Sit down. I want to talk to you. We've had a meeting between George Low and myself and Dr. Robert R. Gilruth. I'm taking a team over to Huntsville today, and what we're going to do is we're going to propose going to the Moon this December." Well, this is in August, so we're talking in four months we're going to go to the Moon. And this sort of catches everybody by surprise here.

He had had a series of these one-on-one-type meetings, and he said, "I want you to go back and tell me tomorrow morning," which would be a Saturday, "tell me tomorrow morning who should work in building up this plan, and I want you to see if there's any reason that we shouldn't fly a mission to the Moon this December." Well, obviously you can pull together a list of reasons you can't do things as long as your arm, but that wasn't the way you did things in those days. What you really did is say, "Why can't we?" So you kept looking for the opportunities that were there.

Next morning I had pulled together a small group of people, myself, Jerry Bostick, and a few others, and [Arnold] Arnie Aldrich, and we went over to see Kraft. He had a very successful meeting at Huntsville that Friday afternoon, came back, and we now had the meeting, which kicked off, at least within JSC [Johnson Space Center], Frank Borman's Apollo 8 mission. We were really told to keep this thing sort of in secret. In fact, we did a lot of these things in those days. You'd try to keep something going, but it was like trying to hide an elephant in your garage. The fact was that all kinds of things had to happen if we were going to go to the Moon.

One of the guys that was really instrumental here was a guy by the name of Bob Ernull. Ernull was one of the young Air Force officers that came to us in the Mercury Program, went back into the Air Force, got his master's program, then came back to Johnson. He still works here today. Bob was one of the first computer gurus, because in those days computers were slow, they were unreliable, big, bulky, etc. He went over to this meeting that we had Saturday with Kraft, and Ernull was given the job to come up with all of the initial mission planning. It's really the question of what day should we launch, what day are we going to return, what ocean are we going to come back to, right down the line.

And Ernull's response was simply, "Chris, if you'll give me every computer in Building 12 and in Building 30 for the entire weekend, I will give you the answer on Monday." It was that kind of a problem. Computers were that bulky. And that's exactly what Kraft did. He turned to Jim Stokes, who was running our facilities at that time, said, "Jim, give him whatever he needs," and Bob went away and literally cranked away at his machines for an entire weekend. I think there were seven computers, these great big mainframe-type computers that were involved at that time, and came out with actually a set of four what we call launch windows.

Three were for the Atlantic, two in December and one in January, and one was for the Pacific in December.

The following Monday, we had a meeting to make this determination, because the Navy, with its aircraft carrier task force, could only cover an Atlantic landing or a Pacific landing, and once it got all set up in that ocean, we were stuck with that for the remainder of the Apollo program. So the decision was made that we'd use the Pacific launch window, and this basically led to the time frame that defined the Christmas Eve mission, and it had nothing to do—we weren't trying to plot it for Christmas Eve. It's just the way the launch window fell out that we would be circling the Moon with Frank Borman on Christmas Eve. But I'm getting ahead of the story.

Anyway, so this mission was now going in the background. We were all trying to get this going, and Kraft wanted to know who should be the flight director, chief of the Flight Control Division. I pretty much nailed down who the flight directors would be, and I gave him Cliff Charlesworth, who was going to be the lead. So that tied Cliff, and again Cliff, Glynn Lunney, and Gerry Griffin for that mission. So this pretty much set the stage. Each flight director—even though we'd be working—was flying missions roughly at two-month intervals during this period of time. So you would finish one mission, you would have maybe an evening off, and then the next day you'd be right back training for the next mission. So we were literally working every day for almost a year and a half throughout this time frame from the Apollo 5 until we got through the Apollo 10 mission . . .

We had all kinds of buzzwords we used as flight controllers in those days, but the discipline was really key, because I don't think there wasn't anybody that wanted to sort of pull the plug on Wally Schirra and leave him to circle the Earth on his own there without any communications for a while, because he was sure ornery. To this day, I don't think anyone has ever figured out what it was that was really the burr under his saddle.

We took a lot of heat from the media as a result of that, but as a result of that mission, we accomplished the checkout of the command and service module, which is what we really set out to do.

We really put that spacecraft through the paces. On the ground, we had a single flight test to do it, and I think this is one of the things that made Wally sort of grumpy. We kept piling a lot of stuff on him, because with only one test to get the job done, every time we saw an opportunity, if a system was working well, we'd go for it, we'd press it, we'd try to get some more testing in there.

On the first flight test of any spacecraft, you're going to find surprises. Remember, this is the first time we've put the manned command service module up in orbit, and you find all kinds of things. Batteries work different in space than they do on Earth. We couldn't quite figure out why we could never complete the battery charging, which is one of the things that made Wally just want to know why the heck we didn't know, and we said, "Wally, we don't know. We're going to have to come back to Earth and test this system and see if we can figure out why it isn't fully charging." So we had a lot of those kinds of issues. We'd make engine burns longer than he had seen before in the pre-mission planning. But the bottom line was that we got the job done and we got Wally home.

We then went into the Apollo 8, the first mission to the Moon. This is one where I was almost glad I was sitting on the sidelines, because I think everybody—Mission Control and the flight directing business—is really amazing. It always seems that the people who are watching the mission get more emotionally involved in the mission than the people who are doing it, because the people who are doing it have got to be steely-eyed missile men, literally. I don't care whether they're 26 years old or 35. The fact is that you've got to stay intensely focused on the job. It is the people who are sitting in the viewing room, I think, who have it the toughest, flight directors who are trying to find a way to plug into somebody's console so they can listen in to what's going on.

But I think that was probably the most magical Christmas Eve I've ever experienced in my life, to actually have participated in a mission, provided the controllers, worked in the initial design and the concept of this really gutsy move, and now to really see that we were the first to the Moon with men. We were at the point where we were setting records, literally, in every mission that we flew in those days, because the Russians had long since ceased to compete, it was obvious that we had the best opportunity for the lunar goal. And this was just a magical Christmas. I mean, you can listen to Borman, Lovell, and Anders reading from the Book of Genesis today, but it's nothing like it was that Christmas. It was literally magic. It made you prickly. You could feel the hairs on your arms rising, and the emotion was just unbelievable.

And was that a total surprise for you to hear that?

Yes. I think it was for everybody. Certain of the crews at certain times just seemed to have a magic ability to select the right thing and do the right thing at the right time, and Apollo 8 was one of those days. I was just happy as all get-out that I was one of the few. Glynn Lunney used a term, that as flight controllers working in Mission Control, to be flight directors or as a member of the team we were always drinking wine before its time, because we were doing things for the first time. We were working missions that people a century from now are going to read about, but we never had time to really savor it, because as soon as we finished one, we'd be on to the next.

As a flight director sitting off-line, not working the mission, this time I did have an opportunity to really savor, to really get emotionally involved with what was going on, where the people working the console never had that chance. I mean, they've got to stay focused, and if they get out of line for even a second, that flight director's going to come down and say, "Okay, everybody get your eye. Get squared away. Get back to business. Let's cut all this crap out." It's interesting to live in that environment. I'll talk about it when we talk about the lunar landing, my feelings about the lunar landing.

Anyway, we went through the Apollo 8 mission, and the next mission was mine. It was with Jim McDivitt, [David R.] Dave Scott, and [Russell L.] Rusty Schweickart.

The Apollo 9 mission was the final mission in Earth orbit, and several things came out of Apollo 9 that were very important. It was my second opportunity to fly a lunar module. I'd flown the unmanned lunar module on Apollo 5, so this put me in a good position for competing with the rest of the flight directors for the lunar landing. We really hadn't established who was going to be doing the landing, and every flight director wanted to be part of the first lunar landing mission, and every flight director wanted to do the landing. So we were all trying to figure out ways to position ourselves so that we'd be the obvious one chosen.

So this gave me an opportunity to really come up to speed now with a manned lunar module, and with two lunar modules under our belt, I not only know the spacecraft but also know the mission control teams that work with them to understand the strategy and the use of the system, to be intimately familiar with not only the systems, but the procedures, the mission roles. So you're really flying missions, but you're also preparing yourselves for the big one.

This was also a good opportunity to really continue this testing of this lunar module, in five days. It was a nine-day mission, and the five days that we were working with the lunar module, we did 10 engine burns, 10 maneuvers, so we did essentially two major maneuvers a day. At the same time, we'd do rendezvous with the spacecraft, actually put the crew in this lunar module, and if we couldn't get the two back together, this crew wasn't coming home, because there's no heat shield on it.

So this led to a series of practices with the command service module where we'd rescue a lunar module. We'd do various rendezvous to demonstrate rescue, etc.

Also, as part of the training—and this turned out to be very important for Apollo 13—part of the training, one day my team didn't do the job right, and when we were debriefing the training, our Sim Supe, which is our training boss, comes to us in the debriefing and says, "Why did you leave the lunar module powered up? Why are we using all that electrical power? Don't you think you should have developed some checklist to power this thing down? Whenever you've got trouble, you ought to find some way to conserve every bit of energy, every bit of resources you've got, because some day you might need it." In debriefing—the guy's name was Jerry Griffith—and we had to say, "Jerry, that's a good idea. We weren't doing our job. We weren't thinking. We were thinking too many other things." So we started developing a series of emergency power-down checklists that really was our first line of defense when Apollo 13 came along.

But I'm getting ahead of the story there, now, because Apollo 9, again, was an incredibly successful mission, the last one in Earth orbit, and it really gave us the confidence that this command and service module and all of the procedures that we had developed so far were going to get the job done.

Now, think about the lunar mission and think about the sequence. We had gone through the sequence of Apollo 7, checking out the command module. Apollo 8, we'd demonstrated our ability to use the Saturn upper-stage S-IVB to inject us out to the Moon. We had proven our ability to navigate in the vicinity of the Moon, to conduct precision maneuvers, to go into orbit around the Moon, and then come back home. Apollo 9, now, we checked out the lunar module, what we call the transposition and docking, because once you get injected to the Moon, you have to separate the command module, turn it around, come back in, and redock with the lunar module, which is still attached to the booster, blow some explosive bolts, and extract that thing. We then demonstrated our ability to perform dock maneuvers which we'd have to do on the Moon.

So we were adding in all of these fundamental building blocks for the lunar mission. The final thing that we had to do was conduct a full-blown dress rehearsal, and that's exactly what we did on Apollo 10. It's to now put all of these pieces together and do a rehearsal for the lunar landing, including making a low pass across the surface of the Moon. This came off incredibly well.

So now we're coming up on Apollo 11. We're getting ready for the lunar landing. As division chief, in the early days shortly after the Apollo 1 fire, I had to run the division. I had all of the controllers and all of the flight directors, and all of the mission planning procedures, the whole nine yards. I also was trying to be a flight director and fly a mission. I made what I guess I'd consider a lucky decision—might consider it fateful—that I could not fly every mission, and what I would do is I would fly alternating missions, and I elected to fly odd-numbered missions. This was very interesting, because it now set up the sequence that put me in interesting positions for the remainder of the Apollo program.

The first of the key odd-numbered missions was Apollo 5, where I learned about the lunar module; Apollo 7, where I learned about the command module; Apollo 9, where I then got more lunar module and command module experience. So by the time it was time to select the flight directors who would work for the first lunar landing, Glynn Lunney had more experience in working in the vicinity of the Moon than I had. I had more experience in the lunar modules. It was very interesting, the balance between the flight directors, because we were all positioning ourselves for what we considered the big one.

But Cliff Charlesworth was the designated lead flight director—and we would always establish one flight director for each mission who would more or less act as the orchestrator for all of the other flight directors. He'd establish what jobs they would do, and we always tried to distribute the workload. The flight director's job requires an incredible amount of knowledge. He has to know the launch vehicle, a three-stage rocket, and all of its systems and the two computers on board that rocket, two spacecrafts with four computers and all of the software in those computers. He has to understand every aspect of the trajectory design of the mission: the launch, the launch aborts, the Earth orbit insertions, various aborts from Earth orbit, translunar injection, aborts from the translunar phase, going into lunar orbit, aborts from that phase, then finally getting into the descent orbit where you get down close to the Moon, the powered descent, the EVA on the surface, the launch from the surface, rendezvous and docking, then back, the trans-Earth injection, and then all the reentry phase. In addition to that, he has to know the science, he has to know the crew physiology.

So we always take and try to break up a mission into digestible chunks, and for the lunar phase of the mission we started adding in a fourth flight director because the job became so immense. Well, the flight directors that were tagged for the Apollo 11 landing, Cliff Charlesworth was the lead, and the lead assigns the mission phase responsibility. So he picked up the launch and the injection to the Moon, and he also picked up the responsibility for the extravehicular operation on the surface. Glynn Lunney had come off the Apollo 7 and the Apollo 10 missions, and he picked up the lunar ascent, getting off the Moon, getting rendezvoused up in lunar orbit. I picked up the responsibility for the landing and the injection from the Moon back to Earth, and then we had Gerry Griffin in there, who picked up the reentry phases.

So, basically we were able to distribute this very large base of knowledge and get it packaged in four flight directors. So we all had a pretty well-balanced workload. I was happy as a clam. The day that Cliff Charlesworth came into the office and said, "You're going to do the lunar landing," I just ricocheted around the office virtually all day, and I don't think my secretaries ever saw me as happy. I had a spectacular secretary, Lois Ransdell, who got me through all of the lunar program. In fact, she's the first woman, and the only woman, who ever became an honorary flight director. She was that well respected by all of the people on the flight control teams.

Now, there's one other honorary flight director, that was [Howard W.] "Bill" Tindall, [Jr.]. Tindall was pretty much the architect for all of the techniques that we used to go down to the surface of the Moon. I think Tindall was probably the single individual who had—if there should have been a lunar plaque left on the Moon from somebody in Mission Control or Flight Control—it should have been for Bill Tindall. Tindall was the guy who put all the pieces together, and all we did is execute them.

I respected Bill so much that when the time came for the lunar landing, the day of the lunar landing, I saw him up in the viewing room, and I told him to come on down and sit in the console with me for the landing. He didn't want to come down, but basically I cleared everybody away and we had Bill Tindall there for landing, and I think that was probably the happiest day of his life. A spectacular guy.

So anyway, those were the two honoraries. Lois Ransdell was Pink Flight and Bill Tindall was Gray Flight, because we always assigned colors to the flight directors. I had the white team. Cliff Charlesworth had green. Lunney had black, Gerry Griffin had gold. So those were the team colors associated with the lunar mission.

Training for the lunar mission was probably the most difficult time of my entire life. You have a training team led by an individual we called Sim Supe, simulation supervisor, and the Sim Supe's

Flight director for the Gemini VII white team, Eugene F. Kranz (left), and George M. Low, Manned Spacecraft Center deputy director, ready a transmission tape from NASA's Gemini VII spacecraft which was received on December 9, 1965. Crew members for the record-breaking 14-day flight of Gemini VII were astronauts Frank Borman, command pilot, and James A. Lovell Jr., pilot. (NASA Photo S-65-61513.)

job is to come up with mission scenarios that are utterly realistic and will train every aspect of the crew and controllers and flight directors' knowledge. It'll test every aspect of the procedures and planning that we have put together. It'll test our facility operators. It'll test our ability to innovate strategies when things start to go bad. Sim Supe has a team of five controllers, and throughout all of the time when we're pulling together the planning for the lunar mission, Sim Supe will sit in our meetings.

If I were having a meeting with Neil Armstrong, Buzz Aldrin, and we couldn't come to an agreement, Sim Supe would write that down. Or at some point we just, in pure frustration, late day, we just decided to cut it off and say, "Well, this is the way we're going to go." So, Sim Supe would log all of those things. He would recognize when we just ran out of time in doing something. He'd realize when we had personality problems in people and in teams. So he would develop training scenarios. He'd look at the mission rules we would write, and he'd say, "Well, I don't think that's the right way to go. My team thinks—but again, they made a decision, so we're going to test them and see if they'll really live with this decision."

Well, Sim Supe is the guy, now, who writes—before the mission, months before the mission, he will take these scenarios, and the training sessions may be only seconds long, some may be minutes long, some hours long, but he's going to write these out, and he's now going to test. Is this really the way we're going to work?

The training process for Apollo 11 began very late, because the lunar landing software in the simulators wasn't ready. So Cliff Charlesworth kicked off the training, and we'd launched Saturn's before, so he basically led off for the first month, in April, and he accomplished the training for the launch phase. So that gave Lunney and me pretty much access to the simulators when we needed to in the months of May and June, because we were going to be launching in July, and we generally cut our training off about two weeks prior to launch because the crew had other things to do, we had other things to do.

So, the first month of training with my team went like a champ. I mean, the training in May, it's almost—we came off missions, we were familiar with the lunar module, we had a hot hand, too cocky. We really thought we were on top of the world. We'd generally run one training session associated with the lunar landing each week. It's hard for me to believe nowadays, watching the mission controller's training, but at the time that we landed on the Moon, we only had about 40 hours of total training in there. We had essentially one week's training in the simulators.

But anyway, the first several training sessions went very well, and then Sim Supe looked at the team and decided we needed to be taught a lesson, and he started increasing the pressure associated with the descent phase. Now, when you're going down to the Moon, just like landing an airplane, there is essentially a dead man's box. No matter what you do, you can throttle up, you can change your attitude, you're going to touch the ground before you start moving back off again, and it's the same kind of condition, but it isn't a neatly drawn line; it's a set of variables, and it depends upon the altitude and the speed at which you're descending down how this box is defined.

If you add in the effect of the lunar time delays, the fact is that everything we're seeing is about three seconds old, and then we have to figure out what our reaction time is and then voice some kind of instruction to the crew on what to do. You have to start defining a set of boundaries. You're going to have to make up your mind before you're actually into the problem. This gets to be very dicey.

We're now in one month prior to launch, and as we're moving into this final month of training, Sim Supe really laid it to us, and it related to this dead man's box and this lunar time delay. We went through a series of scenarios that were almost—it seemed like forever. It was only a couple of weeks, but it seemed a lifetime where we could not do anything right. Everything we would do, we would either wait too long and crash or we would jump the gun and abort when we didn't have to, and the debriefings were absolutely brutal during that period of time.

[Richard] Dick Koos—a bit about Dick Koos as Sim Supe. He was one of the very early pioneers of the Space Task Group. He came in out of the Army Missile Command in Fort Bliss, Texas, because in those days you couldn't hire people with computer degrees. You just went after people who had the experience. Well, the Army was working with computers in their ground-to-air missile program, so he was either a corporal or a sergeant discharged from the Army out of Fort Bliss when they hired him into the Space Task Group, and virtually everybody that we had was hired on a paper basis. There were never any interviews conducted. You just fired in your application, the SF-57, I believe it was, and they'd look at it and say, "Oh, yeah. These people would fit here," and they would just bring you on board, and, well, it was like, "You'll go to training, and you'll go to operations, and you're going to go to engineering," and it was that kind of a sequence.

Well, Koos was one of the guys who ended up in training in the very early days, and training in the early days was incredibly rudimentary, but by the time that we got to the Apollo program, it really had become quite sophisticated. Training in Apollo was about as real—I mean, you would get the sweaty palms, you would have when the pressure was on in a training episode, it no longer was training, it was real, and the same emotions, the same feelings, the same energies, the same adrenaline would flow.

Koos was causing all this to happen, but he decided my team wasn't ready, so he kept beating us up and beating us up and beating us up. Over in the offices where George Low and Chris Kraft would sit, they would listen to our training exercises. They had these little squawk boxes, and they had the air-ground loop, and they had the flight director's loop, and every time we'd have a bad day or a bad session, they'd grit their teeth, until finally Kraft called up one day. It was sort of like this: "Can I help you?" Well, there wasn't any help this guy could give me. I mean, there was nothing. And the only help he could give me was to maintain his confidence that I was going to get it all together.

Behind the flight director's console we have a telephone. Well, what I did is I put a switch in to disable the ringing so that he could call all day long, but I would never hear it. All he would get was a signal that the phone was ringing. But eventually we pulled ourselves out of these pits that we had gotten ourselves into, and at times it got so bad that the prime crew, Apollo 11 crew, Armstrong and Aldrin, just didn't want to train with us anymore, and we didn't want to train with them, to the point where they'd go off in a different simulator and we'd work with the backup crew or work with the Apollo 12 crew.

So it was this kind of fragmentation that you had to pick yourself up and get it all back together, and then somehow, one by one by one, people started picking up, and then pretty soon little groups would have their act together, and this group would act together, and somebody would be in trouble and another group would help them out. I mean, it was just this incredible formation that exists within this team, the intensity, the emotions, the time that we'd spend drinking beer together, the time we'd debrief together, the time we'd go back to our office and kick ourselves because we didn't do as well as we knew we were capable of.

But it finally came together, and it came together just in time, about the time that we were just about ready to finish up the training with the Apollo 11 crew. Then Sim Supe stuck it to us again. The final training runs, invariably, are supposed to be confidence-builders. It's to the point now this is the last time you're going to have an opportunity—generally things are going to go right during the course of the mission, so let's stay within the box, let's build the confidence of this team, etc.

Koos didn't see it that way, and we started our final day of training, and about midway through the day, we had done more aborts, and I was really starting—it was starting to get irritating to me, because what I wanted to do was practice the landing, continue to refine the timing of the landing, but we were boarding when I really felt—and I was really seething, I mean just really frustrated at Sim Supe, but there's no—I mean, he's the boss during training. He's going to call the shots. I think it was either the last or second to the last training exercise—I'm not clear on this and no one is, in fact—but we started off, and midway through the descent training, we saw a series of computer program alarms. We'd never seen these before in training. We'd never studied these before in training. My guidance officer, [Stephen] Steve Bales, looked at the alarms and decided we had to abort.

We aborted, and I was really ready to kill Koos at this time, I was so damned mad. We went into the debriefing, and all I wanted to do is get hold of him at the beer party afterwards and tell him, "This isn't the way we're supposed to train," and in the debriefing we thought we'd done everything right.

Koos comes in to us, and he says, "No, you didn't do everything right. You should not have aborted for those computer program alarms. What you should have done is taken a look at all of the functions. Was the guidance still working? Was the navigation still working? Were you still firing your jets? And ignored those alarms. And only if you see something else wrong with that alarm should you start thinking about aborting." We told him he was full of baloney.

In the meantime, I gave an action to Steve Bales to come up with a set of rules related to program alarms. I want a total expose, and I don't give a damn how long it's going to take him. If he has to work all night or all week or every day from now to the launch, he's going to understand these program alarms.

144

Well, he started off working that evening. I had gone home, and I got a call late at night, and Bales said, "Koos was right. We should not have aborted." They now understood these alarms better, and what they wanted to do was to run another training run the following day, exercising various types of program alarms.

We set up another day training. Now, the Apollo 11 crew had gone, so Apollo 12 crew was the one that we were now working with. We conducted our final day of training, exercising all kinds of various combinations, computer program alarms, right on down the line . . .

We went into the final couple of weeks of training, and the neat thing about the lead flight director is he takes all of the press conferences, etc., etc., so basically it gets to the point where each of the flight directors, except for the lead, has a couple days off so you can really start getting your mind in order, you could get your team in order, you can study the loose ends, you can build the intensity that you need when it comes time to fly.

Each of the controllers does this in an entirely different fashion. I'm an extremely organized guy. There's no way I could cope with the knowledge requirements of the job without total organization. So I have a series of books that I would build. I mean, they're incredibly indexed. I've got every detail you want. I know them by heart. And what I did to identify my books—it would not be allowed in today's days of all the harassment kind of things, but I would take the girls from the *Sports Illustrated* swimsuit edition, and the reason I did this is literally I was frightened to death some day, come mission time, I would have lost one of my books. So everybody knew if they saw a book with the swimsuit edition cover on it, it was Kranz's, and it would find its way back to me.

But anyway, I was super organized, super disciplined, super instructor. Lunney was a soak. I mean, he was so damned smart, it was incredible. That guy had such a gift for being able to assimilate information, knowledge. I mean, it was just unbelievable. Cliff Charlesworth had the reputation—he carried the nickname within Mission Control as the Mississippi Gambler. Cliff—nothing ever seemed to upset that guy. I mean, he was about as loose an individual as you have ever seen in your entire life. It didn't matter if

145

he was going to do something for the first time, it was just no big deal, "Let's just go do it." Cliff was just absolutely almost carefree. He had an ability to—everybody's up here in intensity, and Cliff's way down here, and you wonder if he's ever going to get up for the mission, but that's the way he worked.

Gerry Griffin was, I think, closer—he was sort of a bridge between myself and Lunney. Griffin was very intense, very crisp, but being he was a military fighter pilot like myself, he sort of relied upon structure, procedures, discipline, etc. The days just seemed to—all of a sudden you're there, you're in launch day, and when mission day starts, there's this incredible relief that training is now behind you. There's no more Sim Supe. It's just you and the crew and whatever the heck you were told to do.

There's also—most people look at it as, you know, you're going to be all emotional, you're going to be this, you're going to be hyper. I always find that when time came for a mission, when time came to do something, there was just an incredible degree of just solitude. You just felt so comfortable. I'm trying to find the right word for this thing here, but basically it's just peace. You're at total peace with yourself, and when you reach this total peace, you're ready to go. It's interesting. The adrenaline's pumping, but you have this incredible confidence in your team and in yourself as a result of this training Sim Supe's given you. He's given you your confidence.

It's sort of like the first time you solo an airplane. It's the same doggoned feeling. This instructor has given you this absolute confidence in your ability to get the job done. You never think about, "If I get airborne, I've got to get back to ground. I may crash." It doesn't come across that way. It's just, "Certainly." It's just peace in the business. I used to fly early supersonic fighters, and you'd get out to the end of the runway, and, boy, when you're cruising down the runway at 250 miles an hour and something goes wrong, you've just got to have confidence that you're going to be able to pull it off.

So it's this peace that comes when launch day finally gets there. This changes once you get into the mission because then you've got to build the intensity. But the neat thing was, Charlesworth launched this guy, and traditionally all flight directors show up in there, and you find a place to sit, and you're three or four deep. Every console's three or four deep. Nobody's going to miss the launch for the first lunar landing mission. Nobody.

You got through these days, and fortunately the Apollo 11 mission, like many missions, started off quite easy. Everything was normal. No major challenges. In fact, Lunney, in his log, was tracing what he calls "nits." The flight directors keep a log, and this is the minute-by-minute, second-by-second, blow-by-blow as to what's going on, and then at the end of each shift you pretty much summarize for the next flight director.

You went through this routine the first couple days of the mission, then all of a sudden it's time to change shifts. We go through what they call a "wifferdil," and this is because, unlike Earth, we work on eight-hour shifts in Missions Control, or try to, but all the mission events don't fit neatly into these eight hours. So at times you have to make a time adjustment.

So we finished my last shift before the lunar landing, and I had a 32-hour period until my next shift came up. So now when you do this wifferdil, you have to adjust all of your bodily functions. You've been sleeping here, and now you've got to sleep here. Okay. So not only do you have to do this, your entire team has to do this thing. But, finally, when you're doing this wifferdil, it finally sinks in that this next mission is what the whole program's been about: landing on the Moon. And this is the only time, really, this 32-hour period, and just the first part of this thing, because you're just making time. You're pacing. You've got all this energy, and you don't have anything to do with it. You've got no focus. You can't sleep. Heck, we had six kids, and Marta's trying to figure out some way, "Gene, when are you going to settle down? When are you going to sleep? Are you going to go out to Mission Control Center to sleep? What are you going to do?" and you don't know . . .

I pump myself up each time I get ready to do something. I can hear "Stars and Stripes Forever," by John Philip Sousa. I've got probably thirty or forty records, tapes. Every time you see an opportunity and the kids want to get anything for Christmas or Father's Day or any special event, it's always a new march record.

At this time also we had eight-track players, so I had them in the car. Every place I'd go, I'd have John Philip Sousa. And this is the way I get up to speed, get the energy, get the adrenaline flowing. Driving to work, it's early in the morning, and the drive—there's times when you drive—and I don't know whether it's tired or preoccupation or whatever it is, you would leave someplace and you'd drive to the next and never remember going through League City and the stoplights and it's just—you just hope to God. You knew somebody was looking out for you, because you sure weren't looking out for yourself, and you wonder how you ever got there.

We'd always park behind Mission Control, and the guards are out there, the first time that you know something's different this day is you have roving guards out in the parking lot, which is rare. I remember this one because one of my favorite guards was a guy by the name of Moody. He was a short African American, sharp, crisp, military, I mean starched, impeccably starched, gold tooth in there, and the most effervescent smile you've ever seen in your entire life. And he sort of greets you. And always he would greet you—he knew every flight controller. He knew every detail of every mission. He knew what was going on in every mission. He had exactly the right words to say to every person that would arrive. This was true of the entire guard force, because I think, as opposed to many other managers, people around JSC, these people felt very close to the flight control team.

So Moody gave me the highball, and it was, "You're going to do it today, Mr. Kranz" . . . And you'd highball, and he would smile, and he just set you right. You'd walk in from the parking lot down the pea gravel concrete down there. Walking into Mission Control has always been a favorite of military history. I've read, I think, stories of every major military leader from Julius Caesar all the way up to [Norman] Schwartzkopf, and [George] Patton has always been my favorite, because Patton felt that he was—he had been in the battlefields of Thermopoly, he had been with the Roman Legions, he'd been fighting at Sparta, he had this feeling of predestination. Well, I've always had the same feeling. It's sort of weird. But basically you walk down this hall in Mission Control, and again, I'm not thinking of lunar landing, I just feel that myself and the team I've got, from the time that we were born, we were meant for this day. And it's funny how these things feel.

I'll tell you a bit about the team I've got. Incredible array of people. [Robert] Bob Carlton. Bob is about as dry, laconic—I think he comes from the Carolinas. This guy, if you listen to this tape of the lunar landing, Carlton, it's like he's out picking cotton. I mean, he is absolutely unperturbed. There is nothing—and yet the thing that's interesting in the tape, Carlton is counting down to seconds of fuel remaining. He's telling me we've got 60 seconds of fuel, we've got 45 seconds of fuel, we've got 30 seconds of fuel, and he's right at the point where he's going to have to start a countdown, 15 seconds, we're running out of fuel, and he's just like an everyday occurrence.

[Donald R.] Don Puddy is my lunar module guy. He's very quick. He'll become flight director. Very gifted, very quick. He's basically got the life support systems, he's got the electrical power systems, he's got the pyrotechnics, these kind of things, also got communications. A college graduate, came in straight out of school, tall, lanky, sort of a self-appointed team leader for the people in the lunar module.

Down in the trench I've got Jay Green, who's a Brooklyner, and he's got the Brooklyn accent that, I mean, it almost drenches you with this thing here. You feel you're walking the streets in the Northeast. And cocky. He is so cocky, it's incredible.

Sitting right next to him, on his left, is [Charles] Chuck Deiterich, who's a Texan, great big brush mustache, right on down the line with the drawl. Chuck is retro. He's also the guy that prepares a bunch of messages for them.

To Green's right is Steve Bales. He's one of the original computer nerds. I mean, he looks like one. He's got these big owlish-type plastic-rimmed glasses you got in there. I don't think any of them—they all look like they never needed to shave. I mean, they're baby-faced kind of people.

Cap Com is [Charles M.] Charlie Duke [Jr.], and he's probably the best of the best of the best from a standpoint of the astronauts. He was personally requested by Neil Armstrong to be the Cap Com for this mission, and you've got to respect Armstrong. You've got to respect Slayton, because Slayton also has to concur on this thing. And Charlie Duke was just absolutely a master of timing. It seemed when we were in the pits, Duke always had the right words to say to just pull this team up and convince this team that we'd get it together. I'm sure he had to pump up the astronauts, because they were getting pretty discouraged at times.

And then on my left-hand side I've got a guy by the name of Chuck Lewis. Chuck Lewis is one of my flight controllers who grew up—I hired him fresh out of college. He was on board, like me, only a couple of weeks and he was sent into the heart of Africa at the start of an uprising when they had incredible rioting, the natives in Zanzibar, the ethnic groups, were fighting each other for freedom. The British were getting ready to turn them loose, and the controllers would be walking through town or out in the town, and they'd throw bloody roosters, decapitate them, and throw them at them. This was the symbol of the Afro-Shirazi Party.[1] Lewis also, like in the movies, you'd come back at night, and they were told, "If you have any roadblocks, you see anybody there that's not in uniform, just keep going." Well, this is sort of tough to do in a little Volkswagen in there, and the people have got machine guns.

But anyway, this is the guy who's on my left side. He's pretty much my conscience, my assistant flight director. He's looking at what am I doing and the pace I'm doing it, and any overload I've got, he's going to have to pick up. He's my wing man.

Over to the left is [Edward] Ed Fendell and [Richard] Dick Brown, and we're in the process of transferring communications responsibilities from the systems guys, because it's just too much work, over into a new console position. So these are the key players we've got. Kraft is sitting behind me, along with Gilruth, and I believe George Low was up there.

I sort of left you walking down the hall with George Patton in my mind. You walk into the room, and in Mission Control, when you walk into the room, it's sort of like you're getting the feeling for what's going on. You can feel the atmosphere immediately. You can look in the room, and if you see clusters of people around, we've got a few problems they're working on, but they're all pretty much distributed along the console. Everything is pretty much ops-normal. This room is bathed in this blue-gray light that you get from the screen, so it's sort of almost like you see in the movies kind of thing. You hang your coat up, and everybody had to wear—I don't know—Kraft always had everybody wear coats, suits, or sports jackets, and the first thing you do in Mission Control is take the thing off. So you've got these coats hanging up behind there. I carry in my landing vest, because the vests started a tradition with the white team that goes all the way back to Gemini, and today—this was always a surprise for my controllers—Marta had made me a silver and white brocade vest, very fine silver thread running through this thing. That's in a plastic bag. It's turned inside out because this is always a surprise for the controllers.

Then the rest of the room's atmosphere, it's the smell of the room, and you can tell people have been in there for a long period of time. There's enough stale pizza hanging around and stale sandwiches and the wastebaskets are full. You can smell the coffee that's been burned into the hot plate in there. But you also get this feeling that this is a place where something's going to happen. I mean, this is a place sort of like the docks where Columbus left, you know, when he sailed off to America or on the beaches when he came on landing.

So it's a place where you know something is going to happen. You feel the energy of the room, because, as you walk in, you pass little groups where there's little buzzes of conversation, and you don't waste too many words in Mission Control. You speak in funny syllables, in acronyms and short, brief sentences, and sometimes you use call signs, other times you use first names. It depends upon what the mood of the room is.

I went up to the console. Marta always makes me a lunch. I sort of eat my way through a shift. I think this is the way I show nerves. I have enough food to last—I could be there a week and still have food in there. Generally it's two or three or four of everything—candy bars, food, vegetables, sandwiches. I mean, this bag literally bursts. Put this in there and put a couple cans of soda in the refrigerator we had out in the hallway, continue on out and talk to the SPAN.[2]

SPAN is where we've got the engineering representatives from our contractors, and this is another good place to get the pulse of the room. There are things going right. You talk to the Tom Kellys, principal designer of the lunar modules, or the Dale Myerses. He's in this room. Here you've got the president of North American, vice president of North American. You know, these aren't ordinary, run-of-the-mill people. These are the people that they ought to write more books about. So they're all in there. My controllers are in the other half of the room, and these people don't seem particularly uptight. It's really amazing. You get the feel that, gee whiz, this is just almost a normal day in Mission Control.

You go back to the console and find out it's been more normal than you'd ever expect, because you read the log, and you say, "Glynn, did anything happen?"

He says, "No." And he writes it right in the log. He says, "All we've been doing is chasing nits." Out of a couple hundred thousand pounds total mass of the spacecraft, 40,000 pounds in the spacecraft, he's trying to track down—they can't account for 150 pounds in the lunar module. Well, this is academic when it comes to performing maneuvers, but they have a weight discontinuity. So he's been pursuing this all night long, and it's sort of a distraction. You know, when everything's normal, you're very tuned to pick up anything that happens, so basically you just always stay busy. You're going to stay alert, let your team know that you're a perfectionist, you're after the details.

So, anyway, handover from Glynn, and it still doesn't sink in that today's real time, this is really it. And put on the vest. This has sort of been tradition in there, and the team sees it, and the reason it's always got white is because I'm the white team. Red, white, and blue are the first three flight directors. I've got the white team, and this is sort of a way to get the team loose, to get them, again, a bit relaxed, because you don't want people who are trigger-happy in this business.

We go through the first couple of orbits, and the crew's ahead of the timeline by at least a half hour, and things are really percolating along, no anomalies. It's almost like a simulation. There's many times during this day when the thought would come to mind, it's like a simulation.

Finally we get down to the point now where it is time to finish. We get the crew and both spacecraft timed, [unclear] the spacecraft, and now, instead of being a half hour ahead, we've got to be right on the timeline. There's no more getting ahead of this time-line. We're in our final orbit around the Moon, which is two hours, and roughly for about an hour-twenty, we see the crew, and then forty minutes they're out of sight for us. We're into the final orbit. The crew goes behind the Moon.

There are certain things in Mission Control, and two of them happened, one now and then one later on, that really now indi-cated that this was not a normal day, or not a normal simulation day. The first one—and this was one of the triggering events—the spacecraft is now behind the Moon, and the control team, the adrenaline, I mean, just really was—no matter how you tried to hide it, the fact is that you were really starting to pump. It seems that every controller has a common set of characteristics which is that they've got to go to the bathroom. I mean, it's just to the point where you just need this break. That's all there is to it. It's literally a rush to get to the bathroom. You're standing in line, and for a change, there isn't the normal banter, no jokes, etc. I mean, the level of preoccupation in these people—and these are kids. The average age of my team was twenty-six years old. Basically I'm thirty-six; I'm ten years older. I'm the oldest guy on this entire team.

This preoccupation is the first thing that hits you. All of a sudden, this is different. Then you walk back into the room, and Doug Ward—you hear the voice of the mission commentator, and he talks, and he's commenting that the Mission Control team has returned from their break and they're now going to be in the room through the lunar landing. Immediately that triggers my thought that this team, this day, is either going to land, abort, or crash. Those are the only three alternatives.

So it's really starting to sink in, and I have this feeling I've got to talk to my people. The neat thing about the Mission Control is we have a very private voice loop that is never recorded and never goes anywhere. It's what we call AFD Conference.[3] It was put in there for a very specific set of purposes, because we know that any of the common voice loops can be piped into any of the offices at Johnson. They can be piped into the media, they can be piped into the viewing room, and what we want is an incredibly private loop where we can talk to the controllers when we need to, but in particular it was set up for debriefing, because debriefings are brutal. It was set up for debriefing between the flight control team and Sim Supe.

It also is the loop where, if you've got a flight controller who is errant, you can sit down and say, "Hey, GNC, meet in AFD Conference," and everybody knows that they are not supposed to listen in, but that guy is going to catch hell for something, because you don't chew anybody out in public. It's all a very private affair.

So I called the controllers, told my team, "Okay, all flight controllers, listen up and go over to AFD Conference." And all of a sudden, the people in the viewing room are used to hearing all these people talking, and all of a sudden there's nobody talking anymore. But I had to tell these kids how proud I was of the work that they had done, that on this day, from the time that they were born, they were destined to be here and they're destined to do this job, and it's the best team that has ever been assembled, and today, without a doubt, we are going to write the history books and we're going to be the team that takes an American to the Moon, and that whatever happens on this day, whatever decisions they make, whatever decisions as a team we make, I will always be standing with them, no one's ever going to second-guess us. So that's it.

I finished the discussion and tell the controllers now to return to the flight director's loop where, again, we continue our business. I didn't think about it at this time. After the fact, Steve Bales told it, the next thing I do is I have the doors of Mission Control locked, and I never really realized—we do this for all critical mission phases, but I never really realized how this sunk in to these young kids, and this was the final thing that sunk in in the controllers that, hey, this is again something different from training, these doors are locked, we are out here now. I mean, we've got a job to do.

Then the thing that sinks in on the rest of the M&O staff is a very similar event; we go to what we call "battle short." Battle short is where we physically block all the main building circuit breakers in there. We would rather burn the building down than let a circuit breaker open inadvertently and cause us a loss of power. So now we're in the doors-locked, battle short condition, and the communications intensity starts building up because we know we're now approaching acquisition of signal of the spacecraft, and this is the time we're all going to have to make some decisions.

In the meantime, one of the mission rules that was most controversial approaching the launch, and it has become a personal mission role for me, is NASA management had to have some kind of a ground rule on how much communications and telemetry we must have in order to allow the mission to continue. Kraft and I had lobbied for a very open loop. The flight director will make this decision, where a lot of people wanted very precise. We have to have it at the time of these events. And what they were concerned about, sort of like a crash recorded on an airplane, if we crashed, we wanted to know why we crashed. Okay. So Kraft and I won. We had this very loose mission role.

This is the only one that really bothers me, because it's a pure judgment call. Everything else is not black and white. But basically we've got telemetry, we've got people working on, but this is the unique call of the flight director. We get acquisition of the spacecraft, and from the time that the spacecraft cracks the hill until the time we're on the surface is about a half-hour long. It's about 18 minutes to look at the spacecraft and the telemetry and then 12 minutes from there to the surface of the Moon from the time that we start the powered descent. Immediately, as soon as we acquire telemetry, we're in trouble because the spacecraft communications are absolutely lousy. We can't communicate to them; they can't communicate to us. The telemetry is very broken. We have to call [Michael] Mike Collins in the command module to relay data down into the lunar module, and immediately this mission role has come into mind because it's decision time, go/no go time.

It just continues, broken, through about the first five minutes after we've acquired the data, but we get enough data so the controllers can make their calls, their decisions. Are we good? Are we properly configured? Are we basically at the point in the procedures where we should be?

We move closer now to what we call the "powered descent go/no go." This is where it's now time to say are we going down to the lunar surface or not. Now, I have one wave-off opportunity, and just one, and if I wave off on this powered descent, then I have one shot in the next rev and then the lunar mission's all over. So you don't squander your go/no go's when you've only got one more shot at it.

Come right up to the go/no go, and we lose all data again. So I delay the go/no go with the team for roughly about forty seconds, had to get a data back briefly, and I make the decision to press on; we're going to go on this one here. So I have my controllers make their go/no go's on the last valid data set that they had. I know it's stale, but the fact is that it's not time to wave off.

So, each of the controllers goes through and assesses his systems right on down the line, and we get a go except for one where we get a qualified go, and that's Steve Bales down at the guidance officer console, because he comes on the loop, and he says, "Flight, we're out on our radial velocity, we're halfway to our abort limits. I don't know what's caused it, but I'm going to keep watching it." So all of a sudden, boom! You've sure got my attention when you say you're halfway to your abort limits.

We didn't know this until after the mission, but the crew had not fully depressed the tunnel between the two spacecrafts. They should have gone down to a vacuum in there, and they weren't. So when they blew the bolts, when they released the latches between the spacecraft, there was a little residual air in there, sort of like popping a cork on a bottle. It gave us velocity separating these two spacecrafts. So now we're moving a little bit faster by the order of fractions of feet per second than we should have at this time. So we don't know it, but this is what's causing the problem. It's now a problem.

In the meantime, we've had an electrical problem show up on board the spacecraft, and we've determined that this is a bad meter that we've got for the AC instrumentation. AC, alternating current, is very important on board the spacecraft, because that powers our gyro's landing radar right on down the line. We're now going to be looking at this from the standpoint of the ground so that Buzz won't have to look after it.

So we keep working through these kinds of things, and we give them the go for a powered descent, and immediately, as soon as we go, we can't even give it to the crew directly; we've got to voice this through Mike Collins down to the spacecraft. All through this time, my mind is really running. Is this enough data to keep going, going, going, going? Because I know what I'm going to do in this role. I'm going to be second-guessed, but that isn't bothering me.

We now get to the point where it's time to start engines. We've got telemetry back again. As soon as the engine starts, we lose it again. This is an incredibly important time to have our telemetry because as soon as we get acceleration, we settle our propellants in the tanks, and now we can measure them, but the problem is, we've missed this point. So now we have to go with what we think are the quantities loaded prelaunch. So we're now back to nominals. Instead of having actuals, we've got our nominals in there. So we're in the process of continuing down.

We've now started down, and Bales calls and he says we're not seeing anymore down-track error. His concern was, was it a guidance problem or was it a navigation problem. The difference is, if it was a guidance problem, it will probably continue to be worse. If it was a navigation problem, it will probably remain constant. Well, now he's seen that this residual error, this radial component he's seeing, has not continued to grow. It's remained constant. Even though this is looking now like it's going to be a go, it's going to cause a problem because it's going to move us down. Instead of being at the landing point we had planned, we're now moving further down range to the toe of our landing footprint, which is very rocky.

So now we're fighting—we've got this new landing area that we're going into, we're fighting the communications, we've got the problem with the communications, and we've got the AC problem that we're now tracking for the crew, and now a new problem creeps into this thing, which is this series of program alarms. There's two types of alarms. These are the exact ones that we blew in the training session on our final training day, twelve-oh-one. Twelve-oh-one is what we call a bail-out type of alarm. It's telling us the computer doesn't have enough time to do all of the jobs that it has to do, and it's now moving into a priority scheme where it's going to fire jets, it's going to do navigation, it's going to provide guidance, but it's basically telling us to do some-

Gemini VIII Flight Directors John D. Hodge (left) and Eugene F. Kranz at their console in the Mission Control Room at the Manned Spacecraft Center during the critical reentry maneuver of the Gemini VIII spacecraft into the Earth's atmosphere which occurred on March 16, 1966. (NASA Photo S-66-24807.)

thing because it's running out of time to accomplish all the functions it should.

The other kind of alarm is what we call a DP00: "do program zero zero," which essentially is the computer's going to halt and wait for further instructions. If this occurs before we're very close to the surface, this is an immediate abort. So what we have to do is we have to prevent this computer from going to these bail-out-type alarms into this DP00-type alarm, and once you get this DP00, it's got to be three strikes in a row where it says, "I'm going to idle, I'm going to idle, I'm really going to idle now."

So, Steve Bales is now starting to fight this kind of a problem. Now, it's sort of like-you've seen people driving cars where they end up with this fork in the road and they don't know which way to go, because at the time we're getting these alarms, we also have another very critical thing that we have to do, which is accept landing radar, because without landing radar, we can't update our knowledge. The only knowledge that the crew has of altitude is that which we gave them from Earth, and this is our best guess based on tracking.

Now, in order to get to the surface of the Moon, they have to get an actual altitude, and the actual altitude could be as much as eight or ten thousand feet off. So what we have to do is we've got to get this landing radar in there to update their knowledge of where they are. Well, when the radar starts seeing data, it is telemetered to the ground. We compare this knowledge of where the radar thinks it is with where we think it is and tell the crew whether to accept this radar or not. We think it's good, accept it, put it in your computer, because once you get it in your computer, you're stuck with it.

So I've got half my team trying to look at this radar and decide to go here. I've got the other half that is finally recognizing the significance of these computer program alarms. It's really miraculous to watch this team go, because we now start working these two incredibly complex problems simultaneously. One part of the team is working this, the other part of the team is doing this, and I'm trying to put all the pieces together with Charlie Duke, who's picking the right fragments of conversation up to send up to the crew.

Bales, in the meantime, is now on the loop with his controller in the back room, who's one of the experts on the MIT Draper Labs-provided guidance system, and he is conversing. Do they see any problems in guidance navigation, control, etc.? We don't see it. See, this is what we didn't do in training. We don't see it, so therefore we're going and this alarm will continue.

So we go through this kind of an exercise at the same time we're accepting this radar data. We tell them we're going the alarms, we tell them to accept radar, go on the alarms, you know, radar's good, getting close—you know, we're continuing to work our way down to the surface. Now, fortunately the communications have improved dramatically. Communications are no longer a concern of mine, but they were for about the first six or eight minutes of our descent. But now we're about four minutes off the surface. Communications are just a dream.

Now things are happening, and this team is incredible. Some person—and we've never been able to identify it in the voice loop—comes up and says, "This is just like a simulation," and everybody relaxes. Here you're fighting problems that are just unbelievable and you keep working your way to the surface.

Now, inside the tanks we don't have a gas gauge like you have in a car, or in an aircraft. Once you get at the point, you have a cylindrical tank that's got a round dome at the top and the bottom. The fuel is sloshing around back and forth in this tank, and you have what they call a "point sensor," and this point sensor says that we have 120 seconds of fuel remaining if we're at a hover throttle setting. This is roughly around 30, 35 percent throttle. But now we're no longer hovering, and we hit this point sensor, and this is the first thing Carlton calls out, and it's just like he— every day he calls out "low level." Well, normally by the time he calls out "low level," we have landed in training, and we're not even close to landing here, but he calls out "low level" just like it's an everyday occurrence.

Now, in his back room he has a controller by the name of Bob Nance, and Bob Nance is looking at a recorder which is tracing out actually the throttle position that Armstrong's using, above hover throttle, below hover throttle, above hover throttle, below hover throttle, and he is mentally integrating how many seconds he is above hover throttle and subtracting that from the minutes below hover throttle, trying to give us a new number for how many seconds of fuel we've got. Nance got so good at this thing in training that he could hit it within ten seconds. Now, this is a guy who's eye-balling fuel remaining, and we're getting ready to call an abort on it.

So we get down to the point—and we know it's tough down there, because the toe of the footprint is really a boulder field, so Armstrong has to pick out a landing site, and he's very close to the surface. Instead of moving slowly horizontal, he's moving very rapidly, and 10 and 15 feet per second, I mean, we've never seen anybody flying it this way in training. Now Carlton calls out "60 seconds," and we're still not close to the surface yet, and now I'm thinking, okay, we've got this last altitude hack from the crew, which is about 150 feet, which now means that we've got to average roughly about three feet per second rate of descent, and I see he's at zero. So I say, "Boy, he's going to really have to let the bottom out of this pretty soon."

So, it's watching this horse race between the calls of the controllers, watching what the crew's doing, and then Charlie Duke comes in. He says, "You know, I think we'd better keep quiet from now on." Everybody in the room—you don't have to say that, because this room is deathly silent except for what is on the voice loops, and we're only listening to Carlton's call, and he's just—the last call was "60 seconds," and his next call, the only word that's going to be said is "30 seconds."

So I advised controllers no more calls, because we're now operating in what we call negative reporting. We're not saying a word to the crew, because they're just busier than hell right now, and the only reason for us to abort is fuel. So Carlton called out 60 seconds. Now he hits 30 seconds. Now we're 30 seconds off the surface of the Moon, and very—I mean, incredibly rapidly I go through the decision process. No matter what happens, I'm not going to call an abort. The crew is close enough to the surface I'm going to let them give it their best shot.

At the same time, the crew identifies they're kicking up some dust, so we know we're close, but we don't know how close because we don't know at what altitude they'd start kicking up the dust, and then we're to the point where we're mentally starting, waiting for the 15-second call, and Carlton was just ready to say, "15 seconds," and then we hear the crew saying, "Contact."

Well, what happens, we have a three-foot-long probe stick underneath each of the landing pads. When one of those touches the lunar surface, it turns on a blue light in the cockpit, and when it turns on that blue light, that's lunar contact, their job is to shut the engine down, and they literally fall the last three feet to the surface of the Moon. So you hear the "lunar contact," and then you hear, "ACA out of Detent." They're in the process of shutting down the engine at the time that Carlton says "15 seconds," and then you hear Carlton come back almost immediately after that 15-seconds call and say, "Engine shutdown," and the crew is now continuing this process of going through the procedures, shutting down the engine.

Now the viewing room is behind me—and this is again one of these other things in training, there's nothing in training that prepared you for that second, because the viewing room behind me starts cheering. Our instructors, which are over in the Sim Supe area, sim room over to the right, they start cheering, but we've got to be cool because we have to now go through all of the shutdown activity, but we have to go through a series of what we call "stay/no stay" decisions, because 40 seconds after we've touched down on the Moon, we have to be ready to lift back off again. And every controller, I think, went through his emotional climax that second. I was so hung up by this cheering coming in from the lunar room that I could not speak, and pure frustration, because I had to get going on the stay/no stay. I just rapped my arm down on the console there, just absolutely frustrated. I broke my pencil, the pencil flies up in the air. Charlie Duke's next to me, and he's looking and wondering what the hell has happened here.

And all of a sudden it hurt so much that I got back on track and started, "Go. Okay. All flight controllers stand by for T-one stay/no stay—" you know, and we went through this, and I think every controller went through his climax at that second.

Then, as soon as we finished that, we had another stay/no stay, T-two stay/no stay we had to do, I think it was 10 or 12 minutes later, and these were opportunities for liftoff and go back up, and immediate rendezvous. Once we went beyond T-two, then we had to go through a T-three. While we're doing all this stuff, Charlie Duke's still talking to the crew and [unclear], you know, "Eagle," you know, you've got a bunch of controllers down here about ready to turn blue. Well, the fact is, I don't think any of us breathed for that last 60 seconds.

You hear Armstrong talk, "Eagle has landed," right on down the line. "Houston, Tranquility Base here. The Eagle has landed." And these are all seeping in, and in the meantime we're just busier than hell doing our stay/no stay kind of stuff. We're in between T-two and T-three, and we use a cryogenic bottle, super critical helium, to pressurize our descent engine. Again, one of the things you can never test, the heat soak-back from the engine and the surface now is raising the pressure in that bottle very dramatically, and

162

now we're wondering if this damned thing's going to explode and what the hell are we going to do about it. The fortunate thing was that they had designed some relief valves. They had a pressure disc in there. If the pressure got so high, it actually blows the disc and the valve, rather than blowing the bottle up.

So we're all sweating this thing out here. We're trying to get everything re-synced for the next lift-off, and it's just time, which is almost two hours, between T-two and T-three stay/no stay, it just goes through incredibly quick. Throughout this whole period of time, except for the instant of hearing the cheering, you never got a chance to really think, "We've landed on the Moon."

We get handed over to Charlesworth's team, and I'm going over to the press conference with Doug [Ward] and it was the first time, actually, you really had the chance to unwind and think about, "Today we really landed on the Moon." It's the goddamnedest thing you'd ever seen in your entire life is that you were right there, you were doing all of these things, but every American went through their thing, and we were only limited to a second where we could really imagine and be happy with what we did. It was an incredible feeling.

Anyway, we had gotten to the point where all I wanted to do was finish the press conference, because the phony thing about setting up this mission was, I don't think any human being alive ever felt that Armstrong and Aldrin would get down to the surface of the Moon, they'd make their report, and then they'd go to sleep, but this was the basic plan. Okay, the crew was supposed to rest and get their sleep before going out and doing their EVA.

Well, Charlesworth and Kraft and Low had already made their mind up, and I think Neil Armstrong had made their mind up that, "Hey, there's no way we're going to go to sleep. What we're going to do is, we're going to take a look at the systems, we're going to go through the stay/no stay. As soon as we've gone through that, then we're going to say, 'Hey, we'd like to do EVA.'"

At the time I was finishing my T-three stay/no stay, I had two teams of controllers. I had [Milton] Milt Windler's team who had come in and Milt was at the point where his team was trying to decide whether to just continue this sleep period, or Cliff

Charlesworth's team, which was, are we going to do the EVA? So this whole control center is almost cast. Everybody's euphoric that we've landed on the Moon, but now it's a question of which team is the one that's going to pick up and do the job.

I came back from the press conference—in fact, let me go back a little bit even before we went down to land. When I came into the control center for the start of the shift, I went to the flight directors then I went into the training room, because I wanted to thank my training team for getting us ready for the lunar landing. And the thing that was anomalous about this was Dick Koos, our Sim Supe, was not there. Everybody was surprised, but everybody was sure that he'd show up in time for the landing.

I came back after the press conference, and Koos was there. What we found out is that in the process of leaving his home and coming to work that morning, he had rolled his brand-new TR3 and had pretty much lost the car, but he still managed to get into Mission Control for the landing. So it was interesting. We were all sort of pushing the edge in the way we lived and the things we did and the hours we worked. And that was really neat, to sit back, "My job's over, and it's up to Charlesworth to do the EVAs, and then it's going to be up to Lunney to get the crew back off the surface, and then I've got to come back on duty to get them on their way back home."

But there is absolutely no question that was a day that you're never going to forget, and it's really interesting to try to write that down in some kind of a fashion where you don't lose your audience, you don't lose the people you're writing to, because the intensity, the quality of the training and the way we were prepared put us in a position to deal with problems that—and this was true not only in [Apollo] 11, it was 12, 13, 14, it was everything that we had learned from the very first Mercury mission until the day we came to be tested.

In one of my book interviews, Arnie Aldrich was talking about Apollo 13, and he says that, "You know, at the time that we were really tested," he says, "all the pieces were in place." He says, "The training had literally achieved perfection. The way we prepared ourselves by learning, by doing, all these battles with the contractors that we're going to do these things, nothing is ever given to us, we're going to have to go out and dig it and assemble it, put the pieces together, if somebody had said, 'Are you ready for this kind of a crisis?' you'd say 'yes,' and they'd say, 'Prove it.'"

There's no way you could ever prove it, and yet in [Apollo] 11 we went through the—I mean, the intensity of these seconds with this young group of people, everything that we'd ever done before prepared us for and we were ready. In [Apollo] 12, we got struck by lightning, and we solved that problem in a couple of minutes, and we were ready. In Apollo 13, we had an oxygen tank explode, and this team was ready. In Apollo 14, we had a solder ball in an abort switch just as we were going to go down to the lunar surface, and we waved off and came up with a software patch and fixed that one. In Apollo 15, we had a solder ball in a different switch. It was trying to ignite the main propulsion engine on the way to the Moon. Apollo 16, we had the problems with the gimbals, which would have compromised the landing.

What was interesting is that once this team had achieved what I'd say—and I say this frankly and somewhat boldly—is that we had achieved literally perfection in this art of crisis management. There was nothing that would have ever stopped us. The interesting thing about the history in the book I'm trying to write, it really talks as much about how we got there as what we did once we got there, because it was getting there which was the tough thing, and it was the "tough and competent" coming out of the very searing experiences we had in Apollo 1. It was the difficulties we had with the crew in Gemini 3, where we had literally a rebellion between ground control and the crew over a petty incident.

But it was each one of these things coming together under the incredibly gifted leadership of guys like Kraft and Low and Mueller that put all of this thing together. So when the time came for Apollo 11, we were ready, and we were ready every time from then on we were to be challenged.

We've read and listened to your transcripts from before. The cohesiveness of your team—the emphasis of that keeps coming out; and we know that the emblem on your shirt, we actually have a [lapel] pin, we know that you wear the—

Yes, Mission Control logo.

Would you give us the background of that and those principles that you all believed in that are featured on the [lapel] pin?

The Mission Control logo is an interesting one. At the conclusion of the Apollo 17 mission, we had established a set of values. You know, I talk to people all over the world now. I talk about leadership, the kinds of people we had, I talk about trust that developed between the team, I talk about the values of this team: commitment, teamwork, discipline, morale, tough, competent, risk, sacrifice. I can quote these terms out to these people.

It was these values that built the chemistry, because these are young people. They've never been tested, they've never been tried before, but it's the chemistry that builds within the team so you know within a second whether a person needs help or not. It's a chemistry that builds intuitive communications. It's chemistry that locks people together when things get tough. It's the trust between controllers, flight directors, and crew and even program management that allows us, when things get tight, to make the seconds count, to pick directions and move off in this direction with only a fraction of a second's thought about it and nobody pulling off in a different direction. So it's this amazing place called Mission Control, which is an incredible leadership laboratory. People talk about pass/fail. Well, this *is* the ultimate pass/fail.

As we were approaching the end of the Apollo program, I was looking for some way to leave a legacy for everything that we had ever learned in Mission Control for the next generation of controllers, as Kraft had left the legacy of the flight director. A flight director's got probably the most interesting job description in history. It's only one sentence long: "A flight director may take any action necessary for crew safety and mission success." That's it. I think in American life today in the military there is no job description that is that simple and so frank and so straightforward, no ambiguity. So, Kraft had basically left that legacy.

I wanted to leave now the legacy of Mission Control, and I was trying to find a way that everything that Kraft and myself and

166

Lunney and Charlesworth and Griffin and [Peter] Frank and Windler—I mean, everything we had learned, everything that we packaged in these first 13 years of space. I was up in the viewing room. This was during Pete Frank's EVA. He ran the EVA for the final missions. I would launch off the Earth and I'd launch off the surface of the Moon.

So I was sitting in the viewing room just watching what was going on, and there's a very gifted artist, [Robert] Bob McCall, and Bob McCall would come into Mission Control, and he had the ability—you had to really avoid being distracted by Bob because he'd sit there, and at times you'd be looking over his shoulder, and he could pen out a sketch in charcoal, pencil sketch, in an instant. He'd be watching the EVA on the big TV we've got, and within literally 60 seconds he'd have a sketch of that EVA or he'd be sketching a controller at a console.

I was watching McCall sketch, and it came to mind, I was military, a fighter pilot, and I had designed the insignia for the 355th Fighter Bomber that was had at Myrtle Beach, and got a gifted outdoors artist to draw a peregrine falcon on the attack, and this is the emblem that we used for our fighter squadron. I decided Mission Control—the crew had their patches, but Mission Control had none. The controllers didn't have one. We would wear the crew's patches during their mission, but we had none for ourselves. So Bob McCall—we were in between EVAs. We had three EVAs in that mission.

We had sort of a coffee shop, and went down to the coffee shop. Started talking to Bob, and I don't think Bob was surprised when I said, "Bob, I'd like you to design an emblem for Mission Control." He had just finished the one for the Apollo 17 crew. This was the end of the program, and I said, "Bob, I'd like you to design a patch for us." So we started talking about what this patch should involve, and I started talking about the characteristics, the individual commitment, the putting yourself on the line, the individual "never surrender." I said, "This I want to represent."

So we selected—in those days it was the form—it's this patch right here. This is the original Mission Control patch (pointing to a photo in a book). It's the form of a rocket in the letter "I" representing the commitment, and if you look very closely at this "I," it's truncated at the bottom and the top here. So you can look at that as a stylized letter "I," which is what we bring to the mission.

Then there is this thing called the team, because the "I" has to become a "we" in order to succeed. So we stole the Sigma from Wally Schirra's Sigma 7 mission, his flight, and these two touch each other, because at any time in the mission we work as a team, but we have individuals who must step forward from their team, make their commitment, lead, come up with the answers, and then after that's done, return to their role as a team member. So these two touch each other, and they're always interleaved in whatever emblem we've got. So this is the "I" and the "we."

Then we had four stars down at the bottom, which talked about the tough lessons that we had learned in the early days of space flight: discipline, which came out of the Gemini 3 mission, where we had a fracas emerge, a petty incident, out at one of the remote sites between an astronaut who was sent out to the site and one of my controllers, and we couldn't agree who was in charge. This led to a series of minor incidents between crews and controllers, and the fact was that we were not in the right mental frame of mind for that mission and it must never happen again. The mission, our objective, must always be dominant in what we're doing.

Then there's the morale, because throughout the entire program we were faced with setbacks, and it was only our belief in ourselves and our ability to survive and solve the problem that would get us through the very difficult times that we knew were involved in space.

President [John F.] Kennedy used some interesting words when he set out the lunar goal: "We choose to go to the Moon. We choose to go in this decade and do other things, not because they are easy, but because they are hard." And he recognized how difficult this challenge was going to be, and it was only the morale, the second star in our emblem, that really kept us going through many difficult times, because blowing up rockets, coming back in, getting your act together, seeing missions fail, seeing our target, our Agena targets, go in the drink on Gemini VI and Gemini IX, and coming with alternatives for those things, morale, believing in ourselves was essential.

There's another element of morale, because our controllers, contractors, at the end of Gemini, had to be with us to the next mission, and then they were out of a job, and you want to talk about a very difficult time for my organization, for the controllers, because all the Apollo controllers had been hired. So it was a very difficult time for our contractors, particularly our Martin contractors at that time.

Then we talked about the tough and competent. "Tough and competent" came out of the Apollo 1 fire. It was basically seared into us. We were tempered in that time frame, "tough" meaning we are forever accountable for what we do or what we fail to do, and that has to be a characteristic of the controllers. And "competent," we can never stop learning, never take anything for granted, never be less than perfect.

So, by the time that we got into the Apollo manned program, these values that we had established as controllers had matured. They had gelled. They then put us to the point where absolutely nothing—nothing—would ever defeat us. If you go through the history of the Apollo programs, that was it. This is a living insignia. We changed the characteristics as our organization continued to grow from Mission Control into Mission Operations. I always would meet with my people at least a couple of times a year, and some of the people that I had in the organization felt they were left out because they didn't work in Mission Control; they worked in Flight Design. So basically what we wanted to do—they worked in the software reconfiguration business. And it was a darned good idea. So, basically we changed this Mission Control to Mission Operations. We changed it [Saturn V rocket] to be the Shuttle.

And then, to make sure, again, since this is living and it continues to live, at the time of the Challenger accident, we had another generation of young people who went through the same very difficult learning process, living through catastrophe and having to emerge from the other side, only this group of young people had not been associated with flight tests, they had never had a mission that had failed before. They had flown 24 straight successful Shuttle missions. We had brought our crew back every time, and all of a sudden they lost a crew. So this, we added in a meteor, a dying star, to represent the risk and the sacrifice to the components.

Basically, this insignia is the sum of all of the knowledge that we have ever had and developed that we believe is essential to success in our business. Our motto at the top of this insignia has never changed. It's been "Res Gesta Par Excellentiam," which is "Achievement Through Excellence," and again, this sort of sets the framework for expectations of what we demand of ourselves and what we demand of the people around us.

Mission Control and the Mission Control Team is probably the most successful long-term risk management organization in American history, maybe world history, because we have gone from 1960 through almost 1999, and as a result of anything directly under our control, we've never lost a man in space or a woman in space. The design flaws of Challenger were beyond our capability. There is nothing that we could have done. We had no options.

So, the emblem is very, very important to us now. One of our flight directors, Pete Frank, translated the values "discipline, morale, tough, and competent now" into a statement in words that we call the foundations of flight control. Now it's called the foundations of Mission Operation. It puts these down very succinctly so that each controller has an opportunity to reflect as he approaches his role and responsibility in each mission and as they're tempted to compromise and maybe look at something as good enough, to go back in and say, "This is what we stand for. This is what we've got to press forward."

So it's an interesting cultural set that was established for the controllers and the control team. It's one that's very successful. I take, and every time I speak, I speak about, I'd say, between 70-100 times a year. I'm on the road frequently, and I use this value statement for grade-school kids all the way through presidents of corporations. This value set is universal, and it's the kind of statement that if you can live by, you can say you conduct yourself professionally.

ENDNOTES

1. NASA occupied a tracking station in Zanzibar which, in 1963, became an independent nation in East Africa. Following its independence, the broad-based and predominantly African Afro-Shirazi Party (ASP) had the majority of the popular vote, but despite this, power was held by a coalition of two parties supported by the British. At this time, there was a growing movement for independence from colonialism and its ties throughout East Africa. Many of the flight controllers sent to this region during a mission undoubtedly encountered hostilities. A revolution shook Zanzibar in 1964 resulting in the ASP's Abeid Karum becoming Prime Minister. Later that year, Karume and Tanganyika's Julius Nyerere signed an Act of Union between Zanzibar and Tanganyika to form the United Republic of Tanzania.

2. SPacecraft ANalysis (SPAN) is a management level room in the MCC that acts as an interface between the flight controllers and the engineering community in the Mission Evaluation Room (MER). The SPAN was staffed by flight controllers, astronauts, and program office personnel.

3. The Mission Operations Control Room (MOCR) had a position called the Assistant Flight Director (AFD) who's console intercom keyset had an intercom "loop" called "AFD Conference." This was the primary intercom loop for the AFD but it was extensively used by the flight director for "private" discussion with the MOCR since for a time, its access was limited to only the MOCR. Eventually, more and more "drops" were added and the privacy was lost.

Arthur J. "Jack" Rogers, Jr., chief of the Facilities Engineering Office, at the Mississippi Test Facility (renamed the Stennis Space Center in 1988). Rogers was instrumental in the design, construction, and activation of the Center and significant in the test operations for the Saturn V. (NASA Photo.)

CHAPTER 7

ARTHUR J. ROGERS, JR.

(1 9 3 2 -   )

*Arthur J. Rogers Jr. was born on July 18, 1932, at Gulfport,
Mississippi. After attending the Gulfport public schools, Mr. Rogers
earned a BS degree in civil engineering from the University of Mississippi.*

*After several years as a structural engineer with the U.S. Army Corp
of Engineers, Mobile District, Mr. Rogers began his career with NASA
in 1960 at the Marshall Space Flight Center in Huntsville, Alabama,
where he worked as a facility project engineer. In 1965, he transferred to
the Mississippi Test Facility (now the John C. Stennis Space Center) to
work as manager of the Facility Engineering Office. Since October 1988,
Mr. Rogers served as director of Center Operations at the John C. Stennis
Space Center until his retirement in the early 1990s.*

*Mr. Rogers is a fellow of the American Society of Civil Engineers
and the Chi Epsilon Honor Society. He has been the recipient of a
number of NASA awards, including the Outstanding Leadership Medal
and the Exceptional Service Medal.*

*Mr. Rogers is married to the former Lois Barber. They have three
children: a daughter, Cindy, and two sons, Gary and David.*

Editor's Note: The following interview has been edited from an original interview with Arthur J. Rogers, Jr. that was conducted by Dr. Charles Bolton and Mr. Steven Patterson. The interview, in Mr. Rogers' office at the Stennis Space Center on October 4, 1991, is part of the Stennis Space Center History Project, in conjunction with the Mississippi Oral History Program of the University of Southern Mississippi.

Was there anything in your background that led you into engineering, which eventually led you to your NASA career?

Well, my father was an engineer. He wasn't trained, college trained. He, as they did back in those days, began in road building. He was always involved in highway engineering and construction. So it kind of flowed through the genes from him to have an interest in that field.

My grandfather on my mother's side was a craftsman. He was a tinsmith. He always wanted some of the family or his grandchildren to be an architect. He had a son that was an engineer, and three grandchildren [that] all became engineers.

You come from a family of engineers then.

I think it was the right thing. I knew it from a very early age, and it was never a big question that I wanted to be an engineer. The only questions came in college when the courses got tough. Was I going to be tough enough to stay with them? [laughter]

You said that when you first began to see the space program that you were very interested in getting involved in it. Maybe you could just tell us about the times, the people, how they viewed the space program. Was it an exciting thing to see this thing develop?

As a kid, I remember some nights playing out and looking at the stars and just wondering, "By golly, I wonder what's past those stars. What's past the stars and even beyond that?" Just thinking about that started my interest. It wasn't deep thinking. Just an interest was there to explore or to be a part of something to explore.

174

When we got close to the activities and seeing the hardware and working on the facilities that were to be, testing and building this hardware [used for the Apollo program], and knew we were indeed going to take things forward, there was real excitement. The time that we got into it, of course, we were getting on the threshold of having the manned flights. Sputnik was up and Explorer I had been put up after the Navy's Vanguard collapsed on the pad. That was a low point for most everybody that was interested in space. But when my work in Huntsville got going, it was just on the edge of the Mercury program.

When I went with NASA in 1960, it was interesting to go through some of the buildings, and you'd see in this area that they were testing and checking out the little capsule that the chimp, I think, Ham the chimpanzee was there. So we saw some of that going on. The seven original astronauts would be in and around because they were checking the progress of hardware that was leading to their launch. Then the Russians put up Yuri Gagarin and that gave a more excited pace to this.

I think it was in the spring of 1961 that Alan Shepard went up. That was a real exciting time for all of us. Then, a year or so later, John Glenn orbited. We huddled around the radios and listened to that and couldn't wait to see some of the little, funny, jittery pictures that came back of the first people being in this weightless experience. The nation was excited about that. Everybody in the program was excited and young.

That's a thing that we look back on now and we say, "Golly, we can't let these kids get in here and do these things and take responsibility for this design, this piece of hardware." We forget that we were all at that time given as much responsibility and authority as you were willing to reach out for and do. We were in our late twenties and early thirties. Now we've gotten older and we think, "Kids that young can't be that smart."

But there was plenty of work and everybody was excited. There wasn't enough time in the day to get it done. It was exciting. Maybe you got depressed a little on Sundays or the weekend because you couldn't be out there doing something. You couldn't wait to get back Monday to pick up the slack and start going.

Something the nation was interested in, and it really got a lot of publicity, and you felt that you were contributing to part of that. That just made every day exciting.

Sounds like it was more than just a job.

I think that's a fair summary of it. It was almost like saying, "We're doing all this and having this fun, and they're giving us money for it to boot." But, of course, you had to have the money to keep the wife and the children happy, part of your life. And they enjoyed it too.

In Huntsville at that time, in the early '60s, it was just a vibrant town. You had the old team of the German scientists that had come over here that were in management and leadership positions. You had a lot of enthusiastic young people that were coming on doing work under their guidance, exchanging their experience, and having failures but mainly having more successes than failures.

Let me move ahead just a little bit to bring in this facility. I understand that you played an inadvertent role, maybe, in the selection of this particular site [the present Stennis Space Center]. Could you tell us a little bit about that?

We were working in Huntsville, and being in the facilities office, you're kind of in the hub of all the planning and the expansion going on under Marshall Space Flight Center's responsibility. Which even at that time still included the Cape, which was called Launch Operations Directorate of Marshall. Now it has spun off and is Kennedy Space Center.

So it was all controlled by Marshall?

At that time, in the early '60s, it was a subset of Marshall Space Flight Center . . .

We were aware that the administrator of NASA, Jim Webb at the time, said, "We ought to be looking to the future as we are starting to go to the Moon and determine where we could bring all our

The barge "Pearl River" transports the first stage booster of the Apollo Saturn V, the S-IC, along the East Pearl River in 1967 in preparation for testing at the Mississippi Test Facility. (NASA Photo 67-1962.)

space activity together like a cosmodrome. Put together a spaceport that had the manufacturing, the testing, the training, the launching, the control, all at one place." So he had a group of engineers to look at this and naturally he turned to the von Braun team to look at where we could find a place to put all of those aspects together. So near the facilities office this group was put together, and it included some of our people. But not me directly. They were looking at places from Corpus Christi, Texas, all the way around to Sea Island, Georgia, all across the coast. Came to pass on a Friday afternoon, much as we are here, the committee was looking at this in their report. They took a break and we were chatting with some of the people that were on it at a coffee break, or having a Coke. I said, "What are ya'll doing?" They said, "We're getting the Corps of Engineers to do some soil investigation south of New Orleans, down there in the marsh, for building this new test facility." Since I grew up in Gulfport and another gentleman that was with me was from Mobile, we said, "What are you going out in the marshes for? That's where the mosquitoes are and you've got hurricanes."

"Oh, it's got to be close to the manufacturing plant, Michoud over there."

So we said, "There's got to be a better place than that." So we broke out a Texaco road map and looked for a better location to meet their need for navigable waterways, to be close to Michoud, to be on high ground and not near a lot of population. That's why they were looking at the marshes. But they could understand that we were saying, "You're going to have storms. There's going to be subsidings, you're going to have to dike it. You're going to have to spend a lot of money trying to protect it. You ought to be looking for someplace else." So we looked at the map and said, "Well, gee, we fished around here, we know the Pearl River is navigable up above Highway 90. Look, here in this area is an area that's colored to be high ground, about 25 or 30 feet. We know there's not too many people here. You ought to look at that."

"By golly," he said—a gentleman named Godfrey Hoffman—and he said, "Well, by golly, we'll take it back in there." So they went back into the committee and the next thing we know they were coming out and making phone calls to little post offices around here: Santa Rosa, Napoleon, Gainesville, getting the population [figures].

Then they, three or four of them, chartered a plane on Saturday and flew down here and drove around the area. Kind of from that point on you began to see reports saying this is a good place to have the test facilities since you're manufacturing the Apollo first stage over in New Orleans at Michoud. Very quickly it went on to be selected, in October of 1961, as the place to do the testing. They announced it and started acquiring the land.

It hadn't been one of the sites they were considering before that time?

No, not really. They had been looking at Eglin, and other Air Force bases, and King Ranch in Texas, and some other islands. Again, some of the area down in Florida where Kennedy Space Center is now, and other areas were getting significant looks.

So anyway, I like to think that we were at the right spot at the right time to kind of focus [attention on south Mississippi]. But then it had to win on its merits that we had pointed out.

Would you be working here and in Huntsville? Would you be working back and forth or were you primarily relocated down in south Mississippi at this point?

Initially it was all in Huntsville. Because you were just doing the design and there was limited construction and work going on here, as the Corps was still acquiring the land. Then, beginning in 1964, there was significant work going on here. So we began to have more time down here on site to be dealing with questions in real time. So we used to commute from Huntsville down here weekly. They set up a charter plane, because there were so many people coming down, that would fly from Huntsville down—wheels up at six o'clock Monday morning and then drop us off at Picayune. We'd get out here and have vehicles and things and spend the week working here. Unless you were out in California reviewing some designs or in St. Louis or something like that. Then fly back Friday evening. Of course, those were pretty good happy hours going back. The plane had an open bar, so sometimes during the two-hour flight going back up there, there was a lot of pain forgotten about what went on back here during the week.

That lasted till about June of '65 when it was decided that they ought to start putting that decision process closer to where the work was. They began to establish the working group, the operations group, and a new group that began to emerge called the Activation Task Force down here, where the three very active parties, design and construction and then the activation would take that construction, put all the electronics in and check it out, integrate it and begin to flow liquids through it and gases and prove it out. Then over to operations, eventually to go ahead and do the testing. So it was obvious you couldn't do that decision making from long distance, so we were moved down here.

I moved down in '65 and was finally officially transferred in the end of '65, just before 1966. So it was a transition down here to get involved.

Would you have been one of the first group of people that became permanent onsite employees?

Well, not really, because the test operations as they call them, Mississippi Test Operations, was building up . . . General Electric was a contractor here, and the first government people were managing that contractor who was beginning to get a few buildings that they had to maintain, to take over some of the construction. The fire station was among the first things that was done. So they began to equip it and began to get their emergency procedures prepared. So Mississippi Test Operations, some 20 or 30 people, began to emerge before we got down here.

Then in '65, because of the activation and the operations, they created a new management structure and brought in the gentleman, Mr. Jack Balch, from Huntsville. Under his direction they began to put together an expanded organization here. That included activation, the old working group which brought the corporate knowledge, if you please, from the engineering and the requirements side, and others, and began to build what was going to actually manage and run the place.

Coming back to Mississippi, home to this area was nice. I was pleased to finally accept the transfer and to still continue a meaningful job here.

What did the development of this facility mean to the Marshall Space Center? Was the Mississippi facility seen as kind of a stepchild?

Well, in the beginning it was a very important part of it, because these stages—we'd tested several of these Apollo stages in Huntsville. I lived eight or 10 miles from the site and it just shook the hell out of the houses. Of course, you had shopping centers and schools and things that were much closer than I was. So it [Huntsville] began to realize that this is good but it isn't like the old days of Redstone and Jupiter where there was a little noise; this thing shook. [laughter] Now you are dealing with seven and a half million pounds of thrust. And another one that was a million and a half.

180

There was plenty of work going on, plenty. All of their test stands were full, so this was just a good place to send the rest of this work and let the contractors prove that their hardware met the specifications. So it was just very much of a marriage together. I mean interrelationships were good. Technical people were down here all the time, and we had good communication links back and forth with them.

After Apollo, then, some difficulties began where we felt we ought to go and where Marshall wanted to go in the future. We were cutting back on the amount of hardware. There wasn't the need for that much testing. So they were having some of their test facilities being empty and not having roles. So they said, "We'll just need to shut down this place and we'll do future work up here. Fill up the home base." So as our manager, Jack Balch, was striving to continue the investment that was created here and the commitment that Stennis made to the people of this area to provide jobs, opportunities, excitement, a place for their people to go off to school, and come back and work. Then it began to be a struggle to keep this alive, and as budgets got tougher, every dollar that went here was a dollar that didn't go to Huntsville. So there began to be a little bit of a strain, and it led to some difficult times I think.

In general, all our work today comes through Marshall. They are the development center for the major booster propulsion systems for the agency and always will be. They have that expertise well-entrenched up there. But we are getting to be recognized as the place with the facilities and the capabilities to really manage the testing. The big contractors were told to prove their hardware-type testing at Stennis. We have come full cycle now to begin to define a good, professional, cooperative arrangement to give plenty of work for both parties. This gives meaningful lines of responsibility. We're not taking over their decision making, but we also want to be able to make decisions as appropriate for our work function at Stennis. That's coming together now under the leadership of Roy Estess [director of Stennis] and Jack Lee[1] who is the director of Marshall in Huntsville.

So anyway, yes, there were times, but I think in any industry when you have diverse locations of it and times get tough, they

An aerial photo showing the three engine test complexes at the Mississippi Test Facility. In the foreground is the B Complex, which was originally used to test the first stage (S-IC) of the Apollo Saturn V rocket. The B Complex is a dual engine test stand (B1 and B2 test stands) that has been modified for single engine tests. This test stand is currently being used for testing of the Space Shuttle Main Engines (SSME) and the Low Cost Boost Technology Program. In the background are the two single engine test stands that make up the A Complex. These two stands were originally used to test the two upper stages of the Apollo Saturn V (S-II and S-IVB). (NASA Photo 91-447.)

always want to keep the home office and give up the diversity, the outlying activities. And that's not always happy when you are out on the end. Particularly when you see you have some nice facilities and some capability and feel you can contribute.

When the facility became operational, I guess in the mid to late '60s, and started the testing for the Saturn, was there a sense of mission among the NASA engineers out here to complete those tests?

Absolutely. Dedication first of all to bring all of this facility together. Here is an extremely highly technical facility, but that includes all the support stuff from the sewer systems and the water systems and the roads and the medical facilities and warehousing, all the way to the control centers with sophisticated instrumentation. With barges carrying liquid hydrogen in the amounts that have never been done before. Here's a liquid that's minus 423 to 427 degrees. The coldest liquid known. A brand new stage that had never been fired before, and it was on the critical path, on the road to the Moon. So we had an awful lot of attention from one end of the country to another about getting this place ready to do this testing and to qualify this second stage vehicle. So 20 [or] 24 hour days, seven days a week was the norm and not the exception at that time.

But on the other side, in three years from really starting construction, like the fence, setting the fence out here, clearing in 1962 time frame, well, I guess it's four years, until spring of 1966. We had the first firing here in four years. All of this area, piney forest and everything, was transformed into a high technical facility. People were brought in, people were recruited and trained, and brought to understand what was going on here by the local population. Sure, there were some people from other areas in here. Boy, it was exciting to lead to that first countdown and have the first one, even though it was just two or three seconds. And that's all that was planned, is just to prove it. To see the first fire and smoke was just great excitement.

Of course, it went on from there to activate the next test stand and then to doing the big first stage testing of Saturn IC, which was a seven and a half million pound thrust vehicle. Just to feel your garments being shaken a mile away, hearing that tremendous noise and saying, "You know, that thing, that explosion is under control. People are sitting over there in a blockhouse and they are controlling that amount of energy release. It's amazing!" So it was just a tingling of excitement all the time. And then to know that you were moving steadily and progressively to achieving the goals that were going to put man on his greatest adventure, to leave this planet and step foot on another planet in the solar system was exciting. Time passed fast.

Talking about working all those hours, all those days, what kind of effect did that have on the families of the NASA engineers, the daily life outside of work?

Well, I think in general the families were caught up in it too. Now, they may not have been out here working or couldn't understand some of the frustrations of some of the things. But the excitement was there too. The whole nation was looking at the Apollo program, the whole nation was from time to time looking at Mississippi in this area. Here was the critical event to do. Or here was the next critical test. So when you were doing a test, it was national news. The families were caught up and understanding what that meant. When they could see the guys come home tired, but still excited and ready to get back the next day, I think they knew just how much it meant for them. It was different. There's bound to be a strain on some of the family situations, just like any military, and they can't stand the awayness so some families fall apart, but others stick together and enjoy the excitement and look forward to the next step. Whatever it might be.

After the last test, the last test of the Apollo, the Saturn rockets, how soon was it before people here started to feel concerns about the future of the facility? Was it almost immediate?

No, it was well before the last test. Of course, like any schedule you can begin to see that two or three years in advance, "Here comes the last one out of manufacturing. Here it's coming down, are we going to go further than the Moon? Are we going to do something else?" We see that there is no planning coming on for the next vehicle. At one time we talked about a 25 or 30 million pound launch vehicle called Nova. So you began two years or so to see that the end of the pipe was running dry. But you are so busy doing what it is.

Then, of course, right after the landing on the Moon, those of us that live down here know what happened almost a month to the day. It was Hurricane Camille. That was real devastation to the area. Our attention was pulled away from doing rocket testing and everything, to helping to heal the Coast. Healing in many cases our own lives, what damage or loss had occurred. For some period of time we were focused away from the main line and working with the various entities on the Coast to get back up. To get the municipal services going and get things back to seminormal, which took a year or more.

Three or four months later we got back to the first testing here and then within a year or so the last one had been tested, and I guess, of course, we were doing a lot of planning for the future even before that. What is down-moding, you know? How far down do we down-mode? We are hearing that we are not giving up in space, but we have not defined what the next opportunity is going to be in space. We are talking now, you know, '70, '71.

So it was kind of tough. Here is a bunch of things where we went down from about 3,700-3,800 people here to 900, I guess; it went down around 900, under a 1,000. And we're mothballing things. Then we began to get political support and to bring in other resident agencies in here, such as the National Data Buoy Office and Coast Guard. And some others—the EPA, began to move here to use the buildings, U.S. Geological Survey began to come in. So we began to have some use until about '72 or '73.

Then the decision was made, "Well, we're going to build a flyable system that can go up and come back repeatedly and be low cost." That was called the Shuttle. The new set of engines to be developed for that had to be tested somewhere. I guess here again was a great deal of tug of war between Marshall and Stennis.

"We can test it up in Huntsville, we don't need to do it down there. It costs too much."

"Yes, but these are newer facilities and it can be done better here."

Plus we have a very strong senator who had made big commitments to the people. One: the technical understanding of the agency, that this was the best place to do it. Newest, most modern facility. Two: it didn't hurt to have good strong political support. So then we got assigned for the Shuttle engine testing. So now we've started up again on modifying test stands and beginning to bring together test crews again to test the Shuttle engine. Which started mid-'70s and is still going on. [laughter] And it will go on to some extent as long as we are flying the Shuttle. All the engines will fly so many times and then they'll have to be pulled off and components taken and refurbished. Then before you fly them again, they'll go back for some hot firing tests.

As long as they're flying the Shuttle, we will have some test stands available, one or two. We have three stands that we are testing Shuttle engines on now, but I feel in the next few years, the program will be smoothed out to where we won't be bringing in new components and trying to improve it. One or two stands might be turned over to us to put another program on.

Let me ask you a few questions about some other things. First, back to Camille. You mentioned Camille. What kind of damage did it do to the facility?

It did very little damage to the facility. It was less than a million dollars worth. That was mainly roofs. Like this building, the wind blew all the gravel off the roof. [laughter] It just peppered the cars parked in the lot, causing a lot of grief.

It blew roofing materials off, or vents, or roll up doors. Nothing substantial. As a matter of fact, let me loop through to another story. When we built the facility here we built some buildings called the Stage Storage Buildings, which were like big hangers. Two bays for the first stage and three bays for the second stage. The idea being when we received the stages from Michoud or from the West Coast—the second stage was built out there and shipped through the Panama Canal. When we received them we would instrument them for testing and get them all ready. But if you had bad weather coming—we knew we have hurricanes—then you could get them off the stand and put them in the hangers where they'd be safe.

Well, as we got in business, by the time you put them in the stands and you put all the tubing and instruments and everything in there, and when we hear a storm is coming I've got to have so much time to get those things out. I can only lift them out when the wind is less than 10 miles an hour, or something like that. I've got so many days to disconnect and then I've got to move them and transport them. When you line all that out, every time there

The second stage booster of the Apollo Saturn V, the S-II, is hoisted into the A-2 test stand at the Mississippi Test Facility in 1967. (NASA Photo 67-703-C.)

is a hurricane that got somewhere beyond Puerto Rico, we had to start disrupting testing. We had done that once and finally we said, "Well, let's don't do that. We think they'll be safe. You can leave them in the stand, we'll clean off things. We'll put pressure on the inside so they won't collapse. And leave them." As a matter of fact, during Camille we had stages on each of the stands out there. Two second stages and a first stage. Of course, we secured the stands and pressurized the bird. And stood back for a day or two.

No major damage was done to the facility. In fact, we housed, I think, something like 7,000 refugees out here. People who were practically homeless afterwards or that were flooded and lost things. We were a refugee center.

Where would you house them?

In here.

In these offices?

Yes, you might have a family in here. But also in the halls in this building and several other buildings. In fact there was a—what was it—a nursing home group almost, you know, just on the leading edge of it, came out here with cots and beds and sick people. We put them in another area. We had many places to set them up, and we were able, a little bit, to try to feed them out of the cafeteria. But that ran out pretty quick.

The other thing that got to be so nice is we have auxiliary power units out here. We can manufacture some of our own electrical power. We can't run the whole place, but we can do what's necessary. We could turn those generators on and we could have air conditioning here. We didn't have power for 30 days or so in many areas of the Coast. So it got to be very difficult to get the people to leave so we could get back to business. We had food here, air conditioning, [laughing] it wasn't too bad. And when they go back home there were trees down and everything. Maybe they didn't have a house or something. Honest to goodness it got to be a little difficult to finally get them to say, "We can't keep this up." In fact, I think it was about two weeks before we finally got the last of the people to kind of get on out of here and let us get back to what our real work was to be.

ENDNOTES

1. James R. Thompson was director of Marshall from September 29, 1986, to July 6, 1989. The current director is Arthur G. Stephenson.

James H. "Harry" Guin, former director of Propulsion Test Operations for the
Mississippi Test Facility (renamed the Stennis Space Center in 1988). (NASA Photo.)

190

CHAPTER 8

JAMES H. "HARRY" GUIN

(1 9 3 9 – 1 9 9 3)

James H. "Harry" Guin was born on August 31, 1939, in Wilsonville, Alabama. As a child, he moved with his family to Birmingham where he attended public schools and graduated from Ensley High School. In 1962, he completed a BS degree in engineering at the University of Alabama.

Mr. Guin began his engineering career at Brookley Air Force Base in Mobile, Alabama. He joined NASA in 1963 and transferred to Marshall Space Flight Center but was soon assigned to the Mississippi Test Facility (MTF) and was on-site during the earliest days of its construction. As construction neared completion he moved into operations and was involved in testing the F-1 and J-2 engines for the Saturn V.

From 1968 to 1978, Mr. Guin was employed in private industry. He managed Center Operations for General Electric at MTF for several years, then worked in the oil industry after joining Global Associates. In 1978, Mr. Guin returned to NASA and MTF (at that point renamed the National Space Technology Laboratories where he again worked with the facilities group.

In 1988, Mr. Guin received the NASA Exceptional Service Medal. At the time of this interview, Mr. Guin was director of Propulsion Test Operations for the Stennis Space Center. Tragically, he was killed in a car accident in 1993.

Editor's Note: The following interview has been edited from an original interview with James H. "Harry" Guin that was conducted by Dr. Charles Bolton. The interview, in Mr. Guin's office at the Stennis Space Center on June 30, 1992, is part of the Stennis Space Center History Project, in conjunction with the Mississippi Oral History Program of the University of Southern Mississippi.

You said that you've always wanted to be an engineer. Was there something in your background that made you want to be an engineer?

No.

Why this desire?

I think it was the fact that my father worked in the steel mills and he was always around engineers. He always felt that they really did have a good job, and they were the ones that established what needed to be done and how. So from the time that he was young he just started talking about that. And I really thought it was great. What really motivated me, I guess, was that feeling that you know, my gosh, it was a good opportunity. I was the first person in the family to ever go to college and here's a great profession, engineering. I always did well in math in school, it was never a problem. It just kind of fit in. All the science oriented type of things were never a problem going through high school and all. So it all worked.

Did you have a similar desire to specifically get involved in the space program or did that just kind of happenstance?

It did. I'll tell you what, I was in Mobile and working at Brookley Air Force Base, where, by the way, Roy Estess [current director of Stennis] was also. As a matter of fact, a number of people ultimately came to Mississippi from Brookley because Brookley closed about two years after I left. At the time that I went, the space program really had not picked up any momentum. They had already started talking about it in the early '60s, but it was not really

The Apollo Saturn V Test Complex. In the foreground is the A-1 test stand with an S-II stage booster being hoisted into the stand in 1967. (NASA Photo 68-437.)

known to the general public exactly what all that meant. You know, President John Kennedy made the announcement, but nobody really understood what all that meant.

One day in the Mobile Sunday paper there was a big two-page discussion on the space program. By that time, I guess, Sputnik had happened and everybody had gotten a little bit jittery. I read for the first time about what all this meant. All young people, as far as I know, are interested in space. And it just absolutely caught my imagination in what it was going to do and what was going to be required in all the locations around the country that was going to have to be brought online in order to make it work. I just got the bug. At that time when I saw that, I went ahead; that was Sunday. I left the following Friday, no, the following Thursday and went straight to Huntsville. Went into Huntsville and went over and interviewed. You know, just went in and said, "I haven't been out of school very long and I want a job with the Space Agency." I remember them asking me a question like, "What can you do?" My answer was, "I can do anything. That's all you have to do is give me something and I can do it." They laughed about that and went ahead and hired me. Of course, there was a lot of people being hired. It was a time when they were increasing the numbers and trying to put together a team and so it was not difficult at that time.

Of course, after that when I came back and I spent a month or two before I actually went on board with them at Marshall. By that time everybody at Brookley was interested and wanted to find a way to get [involved]. But ultimately we ended up with a number of people over here from Brookley Air Force Base.

You said that when you were in Huntsville you worked at the test lab. Is that right? What was your impression of the German scientists that were heading that up there? What do you remember about them?

As a matter of fact the day that I hired in and went to the Marshall Center, I was one of 33 people hired that day, actually signing in; there were 32 Germans and me. They had just gotten off an airplane that had flown in. Most of the Germans had already been here for some length of time. . .

Had they also worked with the German scientists during the war?

Yes, they were highly skilled technician-level people that worked very close tolerances and that type of thing. You know, building the hardware. Most all of those weren't able to come over here earlier and all after the war. But since they were building up numbers and they had to have some level of craftsmen, they got the names of all those that had done such a good job in Germany and brought them over. Probably only a half dozen of those could speak English.

My first impression with the Germans was, you know, here I am surrounded in this country with a bunch of Germans. Really didn't know too very much about it. The only thing that I recall was back during the war they used to have prisons camps in this country. . . Most people didn't know that. One of them was just outside of town where my grandmother and grandfather lived. My grandfather bought and sold timber. He sold it to the government, and the government in a type of payment in kind would do him favors. One of the things they did, they allowed the German prisoners to work on his house. Built his house, actually built it. So I got very close to some of them that worked there and learned a little bit of German. I was only about six years old, but I did know some of the German language. So I was able to communicate somewhat.

By and large the German people, or at least those that I've been associated with, are very serious-minded. They are very goal-oriented. They are focused on what needs to be done. They just didn't seem to have an awful lot of playtime, a lot of time to carry on, kid around and that type of thing. Well, as you know, Americans tend to do an awful lot of that. They are very serious-minded people. Heimberg was that way and he was the director of test lab. His deputy was Bernie Tessman. He was, as I've mentioned to you before, one of the more light-hearted Germans that I've ever run into. He was a little bit more playful, a little bit more American, and could speak, you know, fairly well.

When you moved here there were probably still people moving out, the residents that had been here before. Did you get a chance to meet any of them or know any of them?

Oh, yes.

How did they feel about this, this being moved out?

They were highly agitated and upset. Senator Stennis had come here earlier and talked with all the residents. Got up on a stump one time as I understand it and had a talk with them and said that he knew that moving out of their homesteads that had been in their families for generations was a hard thing to do. But that Mississippians were Americans and they had to be a part of such a great adventure and that type of thing. That they would ultimately be very happy that they were a part of history. So anyhow, he did an outstanding job but there were still numerous of them that were highly agitated.

195

There were two or three things that happened that might be of interest. One, this was an open animal range at Stennis Space Center during the early days. . . A lot of the local residents even as far as Bay St. Louis and Picayune would come to this area and let their pigs loose in an open range and let them go out and find food the best they could. Then at roundup time they would come back in and round them all up. They had to crop their ears; that was registered in the courthouse. And so the law was that all pigs and piglets that were following those that had been registered properly became the property of the land owner. They would come in at roundup time and go all over these areas with hog dogs and round them up and catch them. The problem was that this area is so massive that you couldn't catch them all. It only takes a second generation before they are just wild. Let me tell you something, many were not tame any longer, they were wild.

There were a lot that they hadn't rounded up.

That's right. There was an awful lot that they hadn't rounded up. You know, they grow tusks and they got to be mean. Some of the things that we did, we had alerted all the local papers and news outlets that the last round-up was past time. The land was bought, they had already supposedly come in and taken all their belongings out and had gone. We kept putting in the paper, like on this date, all remaining free range animals would be collected and would be the property of the government. So they would come in and do what they had to do. Well, each time we did that some of the local people would gripe about it. So we kept extending the late round-up date. We extended it four or five times. The extensions lasted at least a year's period. Still there was an awful, awful lot of wild pigs. Some tame were also still in the area.

So finally we set a last date. It came and went and what we were going to do was go out and round these up and put them in pens. Well, when they are wild you have to feed them something, grain or corn something like that, to get that wild taste out of them. What we were going to do was have a luau for the employees here at that time. We only had probably about thirty employees

196

After testing is complete, the first stage booster of the Apollo Saturn V, the S-IC, leaves the test stand en route to the Kennedy Space Center in Florida. (NASA Photo 67-2189.)

at that time. So some of the community members found out about that and they once again griped about that. So we let the pigs go. Finally the time came and went and I think a judge finally said, "This is it. This is the end. No more. You're going to get it."

I think the funny part of it is on three different occasions we went out and had to catch these pigs, pen them in, and then had to let them go. The funniest part was how you catch them. If you've never been on a hog roundup—

I can't say that I have. [Laughter]

…It is absolutely an interesting proposition because you don't shoot them. You go out in this wooded area with hog dogs and run these hogs and piglets all over the doggone place and what you do, you run after the dogs that are running after the hogs. Up and down little hills and around until the hogs tire out. Then they will turn and they will fight. Then the dogs will try to corner them.

197

One would grab an ear and another will grab the tail. When you've got that ear they would stop moving.

Sensitive there.

Very sensitive in the ear. Then you just picked them up and threw them in the back of the truck with gates on the side. And that's how you caught them. So we went through that process about three times and then finally it was all over.

So this would be NASA employees out there on these hog drives?

Yes. We had a security group, a Hancock County security patrol that we had hired under contract that included local people. We had about eight or nine people that worked 24 hours a day and they cruised the area and so forth. They were the ones that had the dogs; they knew all the local people.

Another early danger for employees that was very evident was the number of snakes that were here because a lot of the area was boggy and low. We had an individual that came here, the first employee of Boeing Company, which built the first stage of the Saturn rocket. . . The first Boeing person was also a pretty high level manager that was very close to an expert on snakes and reptiles. . . What he decided to do, was to pen the snakes to educate the new employees that were coming in, many of which had not dealt with the type of snakes that were in this area. They were just—you couldn't believe how many [snakes] there were. And so he was going to cage them and take pictures of them and give all new employees an orientation of poisonous snakes that were in this area. A picture of them so they could see what they looked like and so forth because we didn't want anybody getting bit by them. So we had a snake roundup. Well, let me tell you, about 99 percent of all people are not overly enthralled about dealing with snakes anyhow. This dude, you know, he just loved it. He treated those things like a child. It's amazing because you know I've always been fairly well frightened of snakes and everybody else around me was too. But we got used to them after a while. Kind of like you get used to anything else. During that period of time we caught an awful lot of local snakes and took pictures and distributed them. Most of the snakes were found when we dredged the Pearl River.

For the canals?

For the canals. We dredged to a little harbor first there just below the lock which connects the canal system. When they were dredging that area there was a Cajun crew that operated the dredge boat. They would average killing in that swampy area, something on the order of two or three hundred snakes a day.

Gosh!

And those were only the ones that were immediately around or in the boat, I mean the ones that they would have to do something about. They'd see them out in the water, I mean it was that bad. The hogs, the snakes that were here, and the mosquitoes were probably the three things that I remember about those earlier days.

You may have heard this, this was in the early '60s, '63, I guess it was in the year '63, that it got such national attention. National TV would show New Orleans and how bad the mosquito infestation was and how bad it was across this region. Actually numerous cattle in this area were killed because the swarming mosquitoes would be ingested during breathing and suffocate them.

Suffocated them?

They actually suffocated them. The number that they actually breathed into their nostrils would suffocate them and they would die. That got national attention.

Well, I was the one that had the responsibility to see about correcting that kind of problem. I contacted the people at the Bureau of Public Health in Washington. They sent two entomologists down here. I rented a helicopter and took them up and showed them low-lying areas where there was a lot of standing water. Then they took water samples. In the boggy areas in a square foot, their samples showed that there was about two thousand wigglers, water larva, per square foot. That's how bad they were. They took a mosquito count of those flying and landing on a standing person in a one minute time frame. As quickly as they could they would get a count of two hundred mosquito landings per minute. So it was quite a troublesome time.

We had a labor strike here. Remember I told you about the fence, the first fence, the perimeter fence that was put in here. They had a strike on the crew that was putting that fence in because of the mosquitoes. There was no repellent that they could give them to stop them from being bitten so badly.

So they just walked off?

Walked off the job site. Said this is impossible, can't work. The company kept bringing in different mosquito repellents but nothing really worked. It would last for about ten minutes. Well, it was so hot, this was in the summer, I mean it was 95 degrees in the swamp. For crying out loud, the humidity was awful. In 15 minutes all of the *Off* would be gone. Well, they finally found a way to solve it.

What was that?

You might remember this if you ever have a problem with mosquitoes.

[Laughter] Okay.

The entomologist came in and recommended something to us. Came in and he said, "Go down and get the absolute cheapest, smelliest perfume that you could find. Get some linseed or cottonseed oil. Put that smelly perfume in it, mix it up, and rub that all over you." It worked. It would work for about three hours and then you would have to go back and put some more on you. But the oil that was used was the type that would just soak into your skin pores. That would work better than anything else, and it's the only

200

reason we got a perimeter fence that year because those workers were not going to work anymore, period.

It's amazing that anybody lived back here with those conditions. Between the snakes and the mosquitoes. [Laughter]

They were called salt marsh mosquitoes. You get used to that too. I mean I couldn't stand it when I was growing up even having a mosquito in the room, you know, at night how they buzz around. I couldn't sleep. I had to find the thing to kill it. But you get used to mosquitoes like anything else. I could actually be talking to someone outdoors and feel them land on my clothing or bare skin, kill them and never miss one beat in the discussion. You get used to them.

Apollo 7 MCC Activity-Flight Director Glynn Lunney is seated at his console in the Mission Operations Control Room in the Mission Control Center at Houston on the first day of the Apollo 7 mission, October 11, 1968. (NASA Photo S-68-49299.)

CHAPTER 9

GLYNN S. LUNNEY

(1 9 3 6 –   )

Born in Old Forge, Pennsylvania on November 27, 1936, Glynn S. Lunney grew up with a fascination for flight. Focusing that interest while entering college, Lunney recalled studying aeronautics because "I wanted to work in the field of flight which meant aircraft since the field known today as aerospace was virtually non-existent."

After obtaining a degree in aeronautical engineering from the University of Detroit in 1958, Lunney began his career with NASA upon entering the National Advisory Committee for Aeronautic's (NACA) cooperative training program at the Lewis Research Center in Cleveland, Ohio. There, he worked with a group that used a B-57 bomber to launch small rockets high into the atmosphere to make heat transfer measurements.

Lunney first became aware of the work being done in manned space flight after viewing a preliminary drawing of what later became the Mercury spacecraft. When NACA became the National Aeronautics and Space Administration in 1958, Lunney worked side by side with engineers at NASA's Langley Research Center, in the area of manned space flight, as part of the newly formed Space Task Group (STG) which he joined in September 1959 as an aeronautical research engineer.

One of Lunney's first duties with the STG involved working in the Control Center Simulation Group. Here he was responsible for the preparation and operation of simulated missions used to train flight controllers in the manned Mercury program.

After moving to Houston, Lunney began work at the newly organized Manned Spacecraft Center (MSC), becoming head of the Mission Logic and Computer Hardware Section of the Flight Operations Division (FOD), where he was responsible for establishing and coordinating the FOD flight dynamics requirements in the new Mission Control Center (MCC). One of Lunney's key roles at Houston was in helping establish the MCC mission rules for all flight controllers and crews. The philosophy behind these rules is sound enough that they remain in use at today's mission control center.

Lunney continued taking on progressively larger roles and responsibilities within NASA, including chief of the Flight Dynamics Branch at the Flight Control Division. In 1968, he became chief of the Flight Director's Office, a role he assumed throughout most of the Apollo lunar program. While in this role, Lunney's duties grew to supervise all other flight directors and manage the training of flight control teams. From 1970 to 1972, Lunney also acted as technical assistant for Apollo to the director of Flight Operations.

Lunney became involved in NASA's first cooperative space venture with the former Soviet Union. In 1972, he became manager of the Apollo-Soyuz Test Project (ASTP) where he oversaw the planning, coordination, and integration of all U.S. technical and operational details, as well as the supervision and coordination of the contracts and other elements supporting the program. When Lunney was promoted to manager of the Apollo Spacecraft Office in 1973, he continued with his ASTP responsibilities. Working with the Soviet Union required a great deal of skill in diplomacy and management; under Lunney's direction, the ASTP mission was launched in 1975 and proved to be an international success.

During his long and distinguished career with NASA, Lunney served in a variety of different positions. In 1975, he became manager of the Shuttle Payload Integration and Development Program. During this time, he also served two appointments at NASA Headquarters: from 1976 to 1977 he served as deputy associate administrator for Space

Flight and from 1979 to 1980 he became acting associate administrator for Space Transportation Operations. From 1981 to 1985, Lunney returned to Houston where he became manager of the National Space Transportation System program where he oversaw all Shuttle vehicle systems engineering, design, and integration.

In 1985, Lunney retired from NASA and began work for Rockwell International. After the United Space Alliance (USA) formed in 1995 as NASA's prime contractor for all Space Shuttle operations, Lunney became vice president and program manager for USA in Houston in support of NASA's Space Flight Operations contract.

Editor's note: The following are edited excerpts from two separate interviews conducted with Glynn S. Lunney. Interview #1 was conducted on February 8, 1999, by Carol Butler and assisted by Summer Chick Bergen and Kevin Rusnak. Interview #2 was conducted on February 26, 1999, by Summer Chick Bergen and Kevin Rusnak. Both interviews were conducted as part of the Johnson Space Center Oral History Project.

Interview #1

You mentioned. . . that one of the Saturn V missions, in fact, the one right before Apollo 8, had experienced a variety of difficulties.

Yes. Right.

When the decision was made for Apollo 8 and you were going to use the Saturn V again, did you have any concerns about it?

We did, but we talked about this pogo thing that was causing one engine to shut down, and then the mixed wiring caused another engine to shut down also, and there was a good fix for the center engine just by shutting it down early. The engine testing had gone well. Once we were getting to the point of saying we were going to put people on board, to light the thing and fire it, we felt that since we're taking all the risks, we might as well try to get the best gain that we possibly can from it. You could have used the Saturn V to do an Earth orbital flight, but the rocket was oversized for that, and you wouldn't have obtained a full, complete test of it. As a result, we began to adopt the attitude of going for the mission that the rocket was designed for and take it out to the Moon, which was done on Apollo 8. . . So once we got over the initial problems that we had on 502 [Apollo 6], the unmanned flight, and saw that those things were fixed, it became a matter of getting used to the idea of going to the Moon.

And Apollo 8 was quite successful.

Apollo 8 was great. . . Apollo 8 was kind of like the door opener for the lunar landing mission. I think all the people, certainly in the operations team—the flight crews, I think, didn't feel quite the same way, but for us, all that had to be done to plan and execute the Apollo 8 mission says that we really knew how to do that. We kind of opened the door so that the next couple of flights were test flights. Getting to the lunar landing mission was shorter than it otherwise would have been, but we got there with confidence as a result of Apollo 8.

Looking at Apollo 8 and talking about the risk with the rocket, in hindsight, after having seen Apollo 13, there was some risk with the spacecraft to some degree. . .

Oh, yes. There are a lot of—I'm not sure I could recount them all, but there are a lot of times when things happened that, had they happened in other sequences or under other conditions, would have been really bad, but for the most part, the things that happened were manageable, in the sequence we had them in.

Apollo 13, for example, had it blown up while the lunar module was on the lunar surface, we'd have been stranded without a way to get home. So the fact that it blew up when it did didn't leave us very much margin to get home, but at least it was some margin, because we still had a full-up lunar module to live off of. And had it happened 36 or whatever hours later, we'd have been stuck. We'd have lost the mission, we'd have lost the crew, etc. So there's a variety of things that happened where the sequence turned out to be forgiving, if that's the right term, and the program was able to continue without grinding to a halt.

We were lucky. . . if we hadn't gotten to the Moon as quickly as we did and Apollo 13 happened somewhere in the getting ready to go to the Moon stage, it probably would have engendered another debate about, gee, maybe this is too risky and we should-n't be doing it at all, especially if we'd missed the goal of doing it within the decade. It just would have had a different flavor to the discussion than it did.

Apollo 13 happening after a couple of lunar landing missions made people feel confident that, well, if we fix this problem, we can go back and repeat what we were doing before. All of that wasn't still in front of us. We already had that under our belt as two successful lunar landing missions. If we did not have that, then the terms of reference for the discussion would have been different.

You mentioned the buildup to the lunar landing of Apollo 11. In between Apollo 8 and Apollo 11 were 9 and 10, both critical missions.

Both critical, and 9 was primarily—although, of course, we flew the command service module—was primarily the first manned test of the lunar module, and so people wanted to put the lunar module through all the paces that they could in lower earth orbit, and that's what the Apollo 9 mission was scheduled to do and did. I didn't work on Apollo 9. I was around the Control Center, but I didn't have a planned shift for Apollo 9 because, by that time, I was occupied with Apollo 10. Apollo 10 was another step like that, although it took the lunar module out of earth orbit and we took it all the way to the Moon, and we did everything short of the actual descent phase and the lunar surface phase.

So we had to do all the navigation things having to do with the two vehicles in orbit. We separated them. We approximated the rendezvous sequence that we would have when we lifted off from the Moon. So we got through all of the phases of flight except the actual descent itself, and then, of course, the traverses that were planned for the surface work.

So we took the lunar module to earth orbit, did everything we could with it, took it to the Moon, did everything we could with it, and then, on the third flight, we were ready to commit it to the landing and it worked fine. It worked fine in terms of most of its performance. There were a few problems that people had to work around in order to be sure that it got to landing.

Go into a little more detail with Apollo 10. . .

Apollo 10 was a great flight. I was the lead flight director on it, and it was, you know, do everything except the landing phase, is basically the way the mission design came down. A number of us argued at the time that if we're going to go all that way and do all that, then we ought to go land on the Moon. Probably the staunchest advocate of stopping short of the descent phase was Chris Kraft at the time. He wanted us to have the experience of navigating these two vehicles around the Moon, navigating, knowing where they are and how fast they're going so that you can get them back together. Because there were unknowns associated with flying so close to the lunar surface, because the trajectories would be disturbed by concentrations of mass from whatever hit the Moon and it would change the orbit a little bit, and that doesn't sound like much, but you can't afford to miss very much when you're doing what we were doing.

So we debated that for a while, but after a while we all got satisfied that this was the right thing to do. So we set about to do everything. Tom [Thomas P.] Stafford, Gene [Eugene A.] Cernan, and John [W.] Young were on Apollo 10, and we had a chance to do everything short of the landing on that flight. The flight pretty much went by the book. There were a few funny anomalies where the spacecraft got out of configuration at one time and was kind of spinning up or going in a direction that the crew didn't expect, and Cernan reacted to that, I think, profanely on the air-to-ground, but that got settled down and got the configuration right, and they got that fixed, and things went smoothly from then on.

Basically, Apollo 10 was sort of like the last clearance test for the Apollo 11 lunar landing try, and the flight went well, everything behaved well, and basically the whole system, hardware and people, passed the clearance test that we needed to pass to be sure that we could go land on the Moon on the next one. Adding the descent phase and the lunar surface work was a tremendous amount of additional training, planning, and getting ready for that had to occur with both the flight crews, with the people in the Control Center, and, of course, all the people that plan all these flights.

So in retrospect, Apollo 10 probably could have landed on the Moon, but it was a matter of how much do you bite off at a time, and the way it came out, Apollo 10 was absolutely the right thing to do.

Apollo 11 had been such a big event and was covered extensively by the media, and then by the time of Apollo 13, there was little coverage. . .

Yes, we could see that, because, of course, it would show up in the numbers of people from the media who would come here to follow the flights. It would take the form of the coverage that would occur in the television, newspapers, or whatever. And we had a sense that the coverage was dropping off, and a number of people—people react to that in different ways. A number of people felt like they were disappointed and it should always be the same as it was, for example, for Apollo 11.

My attitude, I think, all along was there's just something natural in this. I mean, people pay attention to things when they believe that they should pay attention to them. Certainly the first lunar landing mission was something in that category. But then when you repeat it once and you're going back to repeat it again and again—and I say repeat it. From the outside that's what it looks like. From the inside you're doing a lot of different things, but from the outside it looks like you're repeating it, then the interest and the anxiety about it probably goes down a little bit.

So, yes, we sensed and saw the decrease, the indicators of decreased attention to the flights, and my reaction was that's normal, that's human nature, and it wasn't anything to be terribly distressed about, although some people were more stressed, certainly, than I was about it. But I think it was unrealistic to believe that the attention that the world focused on Apollo 11 was going to continue to be focused on every subsequent flight. It just isn't like that. So, while others might have been more upset, I was sort of benign about it. I thought that was normal, and I didn't get too upset about it.

Looking at the attention of the world and the media, some people have mentioned before that they were so caught up in the Apollo program or the other programs that they kind of lost touch with what was going on in the outside world, like with Vietnam. . .

Let me talk about that. Because the '60s were such a tremendously

Apollo 11 flight directors pose for a group photo in the Mission Control Center. Pictured left to right, and the shifts that they served during the mission, are (in front and sitting) Clifford E. Charlesworth (Shift 1), Gerald D. Griffin (Shift 1), Eugene F. Kranz (Shift 2), Milton L. Windler (Shift 4), and Glynn S. Lunney (Shift 3). (NASA Photo S-69-39192.)

volatile and kind of a tearing apart environment in the United States, and speaking for myself and, I expect, for other people, we experienced all that, I mean, especially Vietnam. So many young men were being killed and wounded. There was a sense that you didn't have any idea of how long it was going to go on, how bad it was going to continue to be, and so on. It was bad. So, you know, as an American or even as a human being, I was affected by all that and all the other things that were going on in America at the time—civil rights, the assassination, marches, the hippie stuff, the drug stuff started to come on the scene.

I remember the convention that was held in Chicago in 1968 where there was so much mayhem, really, on the streets of Chicago—tear gasses and the police hitting people to control the crowds—and the people were expressing their point of view, mostly about Vietnam, that we were on the wrong track and needed to get out of there. It was a very divisive, terribly emotional kind of issue. That and other things, all those other things were terribly emotional, and we still were in the middle of the cold war. The threat from the Russians was real, and on and on. So there was a lot of emotional things that were upsetting—that's a mild word. I mean, "upsetting" is just too mild to capture it. It was distressing the hell out of people in the country, and it had that kind of effect on me.

The difference, though, that I felt for myself and maybe for those of us in the program was that we had a real focus on significant events in the '60s, and we could do something about it. I think a lot of people were frustrated because, depending on what their interest was or what their main concern was during the '60s, most people couldn't do very much about it. I mean, people protested, and that was, by the way, an activity that eventually had its result, but it was a long time frame, and it was not clear that it was going to have a positive outcome.

So a lot of people, I think, were frustrated because there was nothing that they could personally do to make any of these things that might have been distressing the hell out of them come out okay. I mean, there they were, and events were out of their control, and these things were happening. So there was a lot of loss-of-control frustration, I think, that people had over the things that were going on. And people felt it all to varying degrees, I suppose, but I think most people in America felt it pretty strongly at the time. All these things were occurring. It was a very difficult environment, and it all caused people to be stressed and frustrated, perhaps, at not being able to do anything about it.

At least in our case, and certainly speaking for myself, I always had the sense that we were involved in a significant activity of our time, significant for our country and for our country's position in the world, and we were—kind of—stewards. I was one of the stewards for this program to make it come out right. So we could return to our little island or our little Camelot, or whatever you want to call it, that we had here in the space program and that we especially felt here at the Johnson Space Center, where everybody in this thing worked so closely together. Of course, we worked with the other Centers, too, but it was keenly felt here at the Johnson Space Center in terms of the teamwork and the comradeship, and the reliance that you had to have on other people. So there was a strong sense of community and people working together and pulling in the same direction and so on.

So the frustration that other people had, perhaps, where they couldn't do anything about this inability to control these events, we at least had a set of events that we had some active control that we could apply to, even on a personal basis. We could personally do our best to assure that our part of this national scene was going to go well. I think it gave us a sense, perhaps, of insulation from the emotional fallout from all these other things that were going on, the frustration, the lack of control, the stressing part of it. They were all real, but for me it was a little different, I think, than for most of the population, because we had this major '60s activity that we were involved in, and we could actually go do something about it every day. We'd go to work every day and work on it, and we could do something about it.

So in that sense, I think, we had an outlet that most people probably didn't have to express their feelings and their sense of what they thought ought to be done about conditions in the country. We had this thing we could do, so it kept us together, and it was a little bit like, when we did our thing, we were on a little island and around us were all these terrible thunderstorms and hurricanes and tornadoes and earthquakes, which were the events of the time, both nationally and internationally, but they were kind of violent, and you almost had the feeling that they were cataclysmic, although it turned out that they weren't.

You got a sense that there was an impending blowup of all these things going on, but we were on this little island with all this going on around us, and yet we were able to focus on the stuff that we had to do, and in that sense it gave us something that we could control personally and something that we could go do and contribute to, and we could do it every day.

So for us it was probably a rock that we could hang onto, and it did mitigate, to some extent, at least for me personally, the frustration in the sense of out-of-controlness that most other people must have been suffering from. But it was very real. It was very real and very painful. No matter what point of view anybody would represent on a given subject, it had to be painful for everybody that was in America—maybe not everybody, but everybody who wanted the country to do well and come through this stuff. It was very painful for people. It was very distressing. And they're mild words. I think what they were feeling was a lot stronger than that, and there were many points of view on almost every subject.

But we had our island and our rock that we could go back to, that we could do something about. We felt like we were making our contribution. We could contribute what the program was going to contribute to the country. So it was like a solace of sorts, or a port, or island in the storm that was going on all around us.

So we were in a different condition, I think, than other people in the country, and we benefited from that, I mean, benefited from it in the sense that we had a focus and a way to express ourselves that was constructive.

I think it did make a difference. I have found in my research that after the Apollo 8 mission, a woman sent in a telegram saying, "Thank you for saving 1968."

Yes. 1968 was a violent, difficult year. It was the year of the Chicago thing. It was the year of the Tet Offensive that started the year off. Assassinations. I mean, it was awful stuff. The hippies and the drug thing was going on. Everybody had a reaction to that, pro or con or otherwise. . . And then Apollo 8 ended the year with the Book of Genesis being read from the Moon. It was quite a change. It was a very absolutely different kind of public event than a lot of the previous ones, most of which had been with hurt, pain, and agony wrapped around them. This was an entirely different kind of thing, and we were part of it and felt like we were continuing, and that we had more in front of us yet to do.

That's the other thing that happened to us. Because of the pace of things, we never really had time to stop and just enjoy it or even to reflect very much on it in a broad way or an overall way. We never had time to sit around and talk about it. We were always

Glynn S. Lunney at console during the Apollo 16 mission. Lunney later served as the project director for the Apollo/Soyuz Test Project. (NASA. Photo S-71-16812.)

so involved in a mission and then the next one, and then the next one, that we did not have time to enjoy it, perhaps, as much as we should have, although the enjoyment came from the energy and the adrenaline that was pumping the whole time. But we didn't have time to be very reflective about it, and that's really come, for me, in the last 5 to 10 years. It has been a revisiting of a lot of the events and a lot of people, one thing and another, books and movies and coverage and anniversaries, and so on.

I'm now grandfather to 12 little people. I have a sense of what life's going to be for them and that, as a member of the family, I participated in some way. I'm leaving something for them to have some sense of what I had a chance to be a part of.

Interview #2

As you were moving into the more scientifically focused phases, Apollo was beginning to lose support, or funding. Did that affect you at all at the time?

I guess I could answer that in a couple of different ways. When I was doing this work in the '60s, I never thought about what was going to come afterwards. I had this sense that it was going to go on forever. It's probably a failing or an idealism of youth, I guess, but you sort of think like, God, this is wonderful stuff, and you think it's going to go on forever. . .

On the other hand, this was set up, as you said, as a response to what was considered to be a major threat, and this was seen as a way to demonstrate capability and accomplishment in this field, and we had done that. Within the program, it's probably like any bureaucracy or any enterprise. Once you get to do something, you think you'd like to do it several more times and repeat it and so on. And as a matter of fact, at the time, there wasn't an alternative that was sitting in front of us that says, "Okay, now that we've done that, we're going to go do this." It took a little while to develop what we were going to do next.

So there was a sense of inertia carrying the missions along, and there was a sense in the program by a lot of people that they wanted to continue and fly another one, another two, another five, depending on, you know, whoever you talked to. Some people just thought we ought to fly them indefinitely, I suppose. And certainly that was a very strong opinion in the lunar science community, because, to them we had just gotten over all this operational stuff and we were really getting ready to be able to do the scientific things that they wanted to do, and the 15 [Apollo 15] was the first mission that had an extensive set of scientific capabilities after the initial capabilities of 11, 12, and 14. And I'm sure they would want to have continued it, and they would probably still be exploring the Moon today.

But it was a response to a certain set of circumstances. Once responded to, what we saw was acknowledgment, explicit or otherwise, by both the political system in America and probably by the public, that we had indeed achieved satisfaction of the purpose, or that the response that we undertook satisfied what caused it in the first place. I think the political system was satisfied, and the public was satisfied that we had done that, and that there was a sense that it was costing a considerable amount of money. The percentage, for example, of the Federal budget that NASA took in the peak years was like four percent. Today, and for long periods of time, almost all the time, it's been one or less than one percent of the Federal budget. So, four percent and one percent is a big multiple. . . And there was a sense that, with everything else that was going on in America, priorities needed changing, and that was the political sound bite that became popular to describe the fact that we needed to reorient and couldn't be spending this amount of money or this high a percentage of our Federal budget for this kind of thing. In light of the fact that it had accomplished its original purpose, and the return, although seen as very valuable by the lunar scientists, was not seen as so valuable to be worth that kind of money by the political system. And I think that was probably, although they might not have known exactly what the numbers were, the sense that the public had at the time.

So I think it was natural that it served a purpose, but there wasn't a lot of purpose to be served by just continuing to do the same thing over and over again. And it was disappointing to people, I think, but being realistic about it, I think it was appropriate.

You know, other endeavors were different. For example, the opening up of the New World. There clearly was, once people got over here, a reason to come back. . .

In the case of the Apollo sequence, the reason to go back would have been the satisfaction of scientific understanding, or the gaining of scientific understanding about what's going on. That doesn't have—didn't at the time and doesn't today—as high a priority as some other things, like the motivations that were operative at the time of the opening up of the New World. Not only was there trade, there was competition with the other major European powers and so on. There was a different set of circumstances. . . that's why it's difficult, I think, to get another major new initiative in space started, beyond the Space Station. It isn't obvious what the relevance and what the rationale is for going ahead and doing that, especially since it's very expensive. If it's a very expensive thing, it's hard to convince people that we should go do that to satisfy our curiosity. It's not much the curiosity of the American public, at least as I can read it today, and I suppose they would say, "Yes, I'm curious about that, and if it costs 10 bucks, go find out, but if it costs lots more than that (which it would), then I don't know that I'm all that curious about it today. I'll wait another day. And maybe you can do it cheaper in the future. Technology changes and things will change and so on and so on, so what's the hurry. I can do that later."

And I think that's the kind of valley that the space program is in right now in terms of the surrounding environment in the country. There's no urgency to go do something, and there's nothing very specific or tangible that we can articulate that there's a reason for going to do that. If, for example, we could solve all the pollution problems of energy and do it on the Moon or some other place and microwave it back to the Earth, well, that might have some interest, if we could do it at a price that was competitive or effective.

I think it's difficult to sell exploration for exploration's sake when it has a real, real high price tag on it, and that's the difficulty that the space program is going to face in trying to chart a course beyond the Space Station. It's not clear yet what that rationale and relevancy is going to be, but I think we need to struggle to find it, so that we can continue. In the big scheme of things, man has moved, and he's probably going to continue to move, and it will take a while.

On the other hand, the exploration of the New World took a while. I mean, it went on for decades/centuries, so, you know, this will, too, and it will have periods when it's fast and periods when it's not so fast, and periods when it's searching for what we are going to do next, and what we are going to do next is probably going to be an agenda for NASA. It is today, to some extent, but it will become more of an agenda as the Space Station program begins to mature.

Is there anything today that we've talked about that you wanted to close off points on?

Let's see. I will probably say this a couple of different ways, a couple of different times, but when I look back on it, I mean, I think we would all say the same thing in slightly different words. We loved what we were able to do. We loved the opportunity to be able to do it. We loved the—I don't want to dramatize this about the challenge of it—but we loved the newness and uniqueness of it, and the fact that we were able to participate in it. I loved the privilege of being in the role that I was.

When I go back and talk about what happened in space, I find that, gee, I was kind of in a spot doing something on almost all of the significant events that ever happened in the human space program in the country, one way or another. I didn't get to do them all, but I got to do a lot of them. And you know, it was done by— not only what we did, but the whole program—people all over America. They weren't necessarily the best and the brightest at anything, they were just pretty good and dedicated.

What took it to another level was dedication to doing it very well, and you took a set of people that I guess you'd have to say were average, and they were dedicated to doing something that was really, in its sum, far above average. We were all successful at pulling it off, and I think we all carry a sense that we were involved in something that was much bigger than each of us as individuals. What we ended up doing in Apollo was a lot bigger than the sum of all of us and I think we all articulate it differently perhaps, but we all carry a sense of that kind of a watershed event in history, and in the development of civilization and the race as we knew it.

It remains to be seen what forms this new frontier is going to take in the future, but I think all of us who participated would have confidence and faith that, yes, indeed, we have opened up an entirely new thing that was not a window, perhaps, a dimension that was not available before. And I don't know exactly what the form of exploring and exploiting that window will be in the future, but it will be there and it will be significant for all of us here on the planet.

We loved it. We loved the work, we loved the comradeship, we loved the competition, we loved the sense of doing something that was important to our fellow Americans. We were obsessed with it. But it took a lot of average people and a few extraordinary leaders, and we managed to do big things. Maybe that's a lesson to take away from this history. You really can do significant things and accomplish extraordinary things just by the proper energy and structuring of what you want to do, but it has to be meaningful and relevant to people. I think that was part of what was so right about Apollo, articulated, of course, at the beginning by John F. Kennedy. But it was something the American public seemed to need the most during that period in terms of a response and reassurance about America's role in the emerging frontier of space. So I think we have to be sure that we have that kind of component, and not a purely self-serving interest, in our rationales and attempts to justify new exploration initiatives.

Geneva B. Barnes in 1999. Photo courtesy of NASA.

CHAPTER 10

GENEVA B. BARNES

(1 9 3 3 –   )

Born in Tahlequah, Oklahoma, on June 29, 1933, Geneva B. Barnes was first encouraged to enter into government service during her senior year in high school. It was during this time that her business administration teacher urged Geneva (who likes to be called "Gennie") to apply for a job with the Navy Department in Washington. A civilian recruiting officer visited their school looking for talented stenographers and Gennie, along with her two best friends, took the civil service test and received appointments with the Office of Naval Material. This marked the beginning of a 41-year career of Federal service.

After working with the Office of Naval Material for four years, Gennie accepted a position at the Pentagon in the Office to the Judge Advocate General (JAG). Subsequent career positions include being secretary of the Regional Director of the Washington Regional Office of the Post Office Department (now the U.S. Postal Service).

In 1962, Gennie began her career with the space program as a secretary in the Office of Programs at NASA Headquarters. The following year she moved to the Office of Public Affairs where she worked as secretary for the next eight years. Gennie did much of the "behind-the-scenes" work associated with NASA special events, including White House ceremonies and other astronaut award ceremonies and appearances. During the Apollo missions, she assisted in protocol activities at the Kennedy Space Center for four of the flights, including Apollo 11. It was during this time that Gennie served among a select group of support staff who accompanied the Apollo 11 astronauts and their wives on a whirlwind presidential international goodwill mission following their successful Moon landing. From September 29 through November 5, 1969, project "Giantstep" traveled aboard the vice president's plane visiting 22 countries in 38 days.

Gennie's career at NASA included a brief stint as a public affairs assistant to Neil Armstrong while he was the deputy associate administrator for aeronautics in the Office of Aeronautics and Space Technology. From February 1972 to July 1973, she served as administrative secretary to the associate administrator for Aeronautics and Space Technology. Other positions held include administrative secretary to the assistant administrator for Equal Opportunity Programs (1973); administrative officer, Office of Aeronautics and Space Technology (1973-1980); management analyst, Office of Management Operations (1980); astronaut appearances coordinator, Office of Public Affairs (1980-1984); and astronaut international appearances coordinator (1984-1994).

Since retiring from NASA in 1994, Gennie now spends some of her time doing volunteer work at the White House in the e-mail section of the Presidential Mail Office.

Editor's note: The following are edited excerpts from an interview conducted with Geneva Barnes on March 26, 1999, by Glen Swanson at NASA Headquarters as part of the NASA Johnson Space Center Oral History Project.

How did you become employed with NASA?

I was working in the Postal Service at the time that I became interested in the space program. John Glenn flew his Mercury mission and I stood out in a misty rain on Pennsylvania Avenue watching him and Lyndon B. Johnson drive down Pennsylvania Avenue in the parade that Washington, D.C. gave to welcome him [John Glenn]. I went back into my office and called the NASA personnel office and asked if they were hiring secretaries and they said "yes." I took my application the next morning to the personnel office and I was on the payroll by 10 o'clock. [Laughter] That was in the days when NASA could do that sort of thing.

What was your first function at NASA, what area did you first work in?

My first job was working in the old Office of Programs which was headed by DeMarcus Wyatt and it was part of Dr. Seamans' staff. They were planning the lunar missions, how to get there and what might be found once they got there. He had a staff of engineers whose job it was to work on that. I worked there for a year and then I went to work in Public Affairs for the late Brian Duff, who had just been hired by NASA. He had been a newspaper man and was brought in to head up a section of speech writers, mainly to support the administrator. He also coordinated some of the administrator's public appearances and traveled with him. Later, Brian and his staff began handling the astronaut public appearances for special events and post flight activities. I absorbed a lot of what he was doing by helping him put together these itineraries and speaking commitments.

Now when you say handling public appearances, were you in the decision making process?

No. I was still a secretary. I remained a secretary, although as I worked in that office longer and it changed hands, Brian [Duff] moved on to something else and then Wade St. Clair came in. However, I was working in other functions such as the protocol operations at the Kennedy Space Center for the Apollo launches. We also helped with arrangements, setup, transportation, and that sort of thing for a couple of the astronaut funerals here in Washington when they were brought here to be buried. C.C. Williams was one. I was really branching out and doing more things than strictly secretarial work and that was how I spent most of my career at NASA.

I moved into what they called the professional career series, not that secretaries aren't considered professionals, but I moved into the professional series when I went to work for Neil Armstrong after the Apollo 11 world tour. My job for him was to answer all of his public mail. He had a secretary who handled his public appearance requests and I rarely got involved in that. Mainly my job was to answer the public mail. After he [Neil Armstrong] left the agency I floated around in a couple of other offices as a management analyst and administrative assistant. Then I had an opportunity to go back to my old office where I had started in public affairs and I began working on the Shuttle astronaut schedules. I was hired as an appearance coordinator and set up their appearances after they flew on Shuttle missions. For the earlier crews, their public appearances began after the post-flight debriefings and lasted about a month. I traveled with the first five Shuttle crews to locations in the U.S. and stayed with them until their appearances were over.

After the Apollo 11 crew returned from the Moon you were involved with this world tour. I was wondering if you could share your stories about that.

To my knowledge this was the first time that there had been a world tour for a crew of astronauts. There had been international appearances by astronauts before but to my knowledge, this was the first time that a crew had been sent on a world tour. President

Nixon wanted to send the Apollo 11 crew, the first crew that landed on the Moon, to share information gained from the flight with other nations and to share plans for future space exploration. The State Department and some of the President's staff set about working on this project with NASA. We did a lot of work to get ready for it. We spent a lot of time over at the State Department putting together briefing books and proposed schedules which changed as we went along.

When did they start as far as the initial planning to get this tour going ?

As I recall, we started in early September of 1969 which meant we all had to get our up-to-date shots and we had to get all of this information put together to support the astronauts. The State Department made the arrangements for the Air Force Special Air Missions people to be involved and we used the vice president's plane for the entire trip. It was considered such good duty by the presidential pilots that they split up the task. One group took the first half of the trip and another group met us in Rome and finished the trip.[1]

We left Andrews Air Force Base, on September 29, and went to the Johnson Space Center [JSC] to pick up the astronauts and their wives. Bill Der Bing and Dr. Bill Carpentier were the JSC staff members.[2] Our first official stop was in Mexico City.[3] We did that all in one day. We went from Andrews Air Force Base to Johnson Space Center to Mexico City which was a preview of things to come.

You didn't really have a rehearsal for any of this. The planning for this appeared after the mission was completed. They had not talked about this prior to the actual mission?

You know I don't know at what point they started talking about it. My boss, Wade St. Clair, and Julian Scheer, who was head of Public Affairs for NASA at that time, were the people that were involved in talking to the White House people and the State

Department people. The overall mission director was the deputy chief of protocol for the State Department, the late Nicholas Ruwe. Of course, NASA had to agree to it before they could start any of the planning, but I'm not privy to when it actually started.

Their first trip was to Mexico City and again it was pretty much the astronauts and their wives. Did they have any other friends or relatives accompany them?

No, it was just them and the support people. Mexico City had an airport arrival ceremony involving Mexican government officials and the American Ambassador, followed by a motorcade. That same day, there were civic events, a meeting with the President of Mexico, a news conference, and a reception hosted by the American Ambassador.

The countries that you decided to visit for this trip . . . those were planned in advance? Did you have to make the preparations, the contacts, and so forth well in advance? Were their some countries that you asked to visit and were declined?

The State Department contacted American embassies in the countries being considered to visit. For instance, I believe there was a plan for us to go to Israel and Egypt in the first proposed itinerary. But that was taken off, presumably for political reasons, I don't know, but we didn't go there. The proposed stops were all planned out by the State Department working with American embassies in a lot of locations. I don't know the process they used to take different countries off and add others.

So the trip to Mexico City was the first stop. Did you learn some things from the first stop in anticipation of what was to come?

I think we were sort of overwhelmed and I don't really think we knew what was in store for us. There were large crowds everywhere we stopped, but Mexico City was the first where we were exposed first hand to such large masses of people. There was a lot of inter-

The Apollo 11 astronauts, wearing sombreros and ponchos, are swarmed by thousands as their motorcade is slowed by the enthusiastic crowd in Mexico City, their first stop on the Giantstep—Apollo 11 Presidential Goodwill Tour. (NASA Photo 70-H-1553.)

est in the astronauts even up to the time that we landed there. I believe that there were local people trying to get on their schedule, to get them to appear at various things. There was quite an effort to hold down on the number of events they could do in one day. I believe that the people in the decision making process probably got a taste of what it was going to be like, that is, being pressured to add last minute events. From that viewpoint, we learned a lot of what was about to happen to us . . .

We went on to Bogota [Colombia]. Generally we were overnight in most of these places, but in Rome and Bangkok we spent three nights. Colonel Aldrin left the group in Bogota for a trip to the U.S. for a speech in Atlantic City and rejoined us in Las Palmas [Canary Islands].

Did they stay on the aircraft overnight?

No. The State Department had arranged for a section of the hotel where we were to stay to be blocked off just for us, just for the astronauts, their wives, and the staff people. We had a control center in each one of these little sections. The Embassy staffed it. We had office machines to use, typewriters, copy machines. There we could assemble schedules and type up 3" x 5" cards that the astronauts liked to use with their notes. I would type these up. We also had office space in the back of our aircraft.

We tried to keep up with thank you notes. We started in Mexico City and after we left there Armstrong had some things that he wanted dictated to people that he and his wife, Jan, had met. I did some of those. But as we moved along, we found that there were so many people to thank for so many things that we couldn't keep it up. We ended up just keeping a list of people, their affiliation, and the address. When the trip was over, the State Department and the local embassies prepared appropriate thank yous to send back to the people.

The trip to Brasilia [Brazil] after Bogota [Colombia] was an unexpected stop. The U.S. had suspended diplomatic relations with Bolivia and we weren't allowed to use a more direct route over Bolivia to get from Bogota to Buenos Aires and Rio de Janeiro. So after leaving Bogota, we stopped in Brasilia. We were there an hour and a half for a refueling stop. We got a tour of the city in a bus. The mayor and other representatives from the local government met the astronauts in one of the official buildings. The two official buildings there are shaped like saucers. I couldn't get over that.

Maybe they represented coffee?

You know you're probably right. After all these years I hadn't thought about it. The mayor and his delegation met the astronauts for an official welcome and that was the extent of what we did in Brasilia. It was straight forward and uncomplicated. I think we were two or three days in Rio de Janeiro. From there we went to the Canary Islands and we were already getting tired. People were starting to become ill by then. It just started with one or two people with sort of flu-like symptoms and then as we got into Europe, everybody had been ill, including myself. I became ill in Madrid along with two or three others. There is a picture that somebody had of one of our staff sitting on the curb in Madrid after getting out of the car to wait for a medical person to come and take him to the hotel. When we got to London, Dr. Carpentier went on national television to deny that the astronauts had brought back a lunar sickness and that all the staff were becoming ill as a result of being exposed to this illness.

We were involved with a lot of time changes and trying to avoid local foods that would cause health problems. With that much closeness you eventually got on each others nerves. We were staying together in the same part of the hotel right next door to each other. When we were flying from one place to another we were all confined in an airplane that got to be kind of small . . .

I think that all of us eventually came to look upon the airplane as being home away from home and we were glad to get back to the airplane. The crew would always welcome us like they were totally rested and were ready to help us out with whatever we needed, cooked our meals, and waited on us. We could eat the food, we could drink the water, we could take a nap.

You mentioned that people got on each other's nerves during the time. Were their any incidents that you can recall that stand out?

No, I don't think so. But somebody very cleverly designed a silly thing to break the ice and to help keep everything in proper perspective. They called it the "personality of the day." If you were caught being unkind to somebody or having an attitude problem, you could be assured that you were going to appear in one of these write-ups. These were humorous little things that were passed out once we got on the airplane. Just silly little things that kind of made us laugh. You were tired. You couldn't get proper sleep. It just stood to reason that you were going to be on the edgy side sometimes.

Was there a press group with you that wrote reports and news articles and then transmitted these through the local embassies to the local papers back home?

Yes. We had two USIA [United States Information Agency] advisors and four Voice of America [VOA] staff members traveling with us. One USIA fellow was a writer and one was a photographer that recorded all of the stops. They were filing reports through their agency channels and were giving information to the local press. There were also news conferences at all of the stops. One of the VOA staff members was a motion picture photographer who recorded all the major events in all the cities. Before we left on the trip, there was some discussion about one or two press reporters being allowed to accompany us on the entire trip and flying on the plane with us. But this idea was scrapped.

I remember seeing in one of the photographs that were taken on the tour, perhaps it was in the Canary Islands or it might have been in Zaire, the crew were greeted by a group of local dancers . . .

. . .There was an evening social event in Zaire at the president's palace. There were dancers there and I know that one of the aides to President [J.D.] Mobuto had an evening banquet to which all of the staff were invited to attend. None of them were women, and, as I recall, the theme of the dance was "Stalking the Lion." They were all men but they obviously had just come in from the bush and hadn't bathed in months. [Laughter] That was quite an experience. They were gaily clad in their costumes and feathered hats. There was also a more public evening program where dancers performed. I remember Buzz Aldrin leaping over the guard rail where the astronauts were seated and started dancing with members of the group. Quite a crowd pleaser. The picture was in the local newspaper the next day.

I noticed that gifts were exchanged during these visits.[4] Can you recall any interesting stories about some of the items and exchanges that occurred while these gifts were presented?

Most of the staff were not included in the official state receptions where gifts were exchanged.

When you went to Berlin was that particularly memorable?

The staff was included in the motorcade through the city, en route to city hall and there was a stop for astronauts to visit the wall. We passed a couple of check points and saw a Russian tank at one. The astronauts were accompanied there by the mayor and the U.S. embassy people. The astronauts made brief remarks at the wall, mentioning a young man who had recently been shot leaping to freedom, and they signed an official visitors book.

I don't know how the astronauts kept up with all of this. They were the ones that were on the front lines. They were the ones making the appropriate remarks and the speeches to the heads of state. As if the rigors of the world tour itself weren't enough, two of the astronauts, Colonel Aldrin and Colonel Collins, left the tour to complete appearances elsewhere—one previously mentioned, to Atlantic City by Colonel Aldrin. Colonel and Mrs. Collins had rejoined us in Berlin after a side trip to Genoa, Italy, where he received City of Genoa Colombiana Medals and an International Institute of Research award. We were doing our best to keep up with it. As we traveled we would get cables from cities like two or three stops ahead of us that would have a proposed final schedule. One of the jobs that I did was to go through the cables and pick out what was being planned ahead at the next stops. Then Wade St. Clair and Nick Ruwe would review and give a thumbs up or thumbs down to what was being proposed.

We were always saying that we needed a rest stop, but we had one true rest stop. That was in Rome. The American ambassador post was vacant at the time of our visit and the embassy was opened to us for the afternoon. We used the swimming pools and the tennis courts and just lounged around and ate American hamburgers, hot dogs, and potato salad. Some of us toured the catacombs under the embassy grounds. That was a true rest stop. But we had another, what was called a rest stop in Belgrade [Yugoslavia] where the astronauts started off on a duck hunt with the representative of President Tito. I believe it was his deputy prime minister. The astronauts' wives were taken on a hydrofoil trip down the Danube and the rest of us were put in cars and buses, including the Air Force crew and all of the support crew, and were taken on a driving tour of the countryside. We all ended up for lunch at a country lodge and we were served a seven course Serbian lunch including roast pig and slivovitz. Have you heard of slivovitz?

No.

Its the national drink. It's served in little tiny glasses and burns as it goes down. It's like fire water. One sip is sufficient, just to say you've experienced it.

That luncheon lasted until four or five o'clock in the afternoon. As we were leaving the lodge, my boss told me that he had gotten word that the chiefs at the hotel had obtained the ducks that the astronauts shot and were dressing them for dinner. After having all of that food for lunch we had to attend the duck dinner. And we couldn't say no lest we offend our hosts. Most of the traveling party appeared for dinner. That entire day was an experience I'll never forget.

What was the longest flight or duration that you had between stops?

There were two flights. The first was from Bangkok [Thailand] to Sydney [Australia] all in one day with a two-hour stop in Perth [Australia] where the astronauts were in a motorcade and made remarks at an official greeting hosted by the city officials. We started out early in the morning from Bangkok and landed in Sydney around midnight on the same day. The Prime Minister of Australia met us there. The other flight that comes to mind was from Tokyo to Andrews Air Force Base with a refueling stop in Elmendorf Air Force Base near Anchorage, Alaska. Originally, the plan was that we were to be given a two-day rest in Hawaii before flying back home after we were finished in Tokyo. But President

234

The Apollo 11 astronauts and their wives (far right) are welcomed to Australia and Sydney on their arrival at Kingsford-Smith Airport by the Prime Minister John G. Gorton, M.P. The presidential aircraft that carried the astronauts, their wives, and support crew during the world tour is shown in the background. (NASA Photo 70-H-1594.)

Nixon wanted to take his family on a vacation and their plans were to start fairly soon so we had to fly straight from Tokyo to Andrews Air Force Base in one leg. That was quite a tiring experience . . .

One of the more memorable experiences we had was in Dacca [East Pakistan]. As soon as we approached the airfield you could see all of these people at the airport waiting to see the astronauts. The crew had to shut off the engines as soon as we landed because the crowd broke through the restraints and came running out onto the airfield. Since we could not approach the terminal, cars were sent out to the plane to take us on the motorcade to the hotel. It was a hot day. A very hot day.

When you drive in a motorcade you are supposed to keep almost bumper to bumper to keep people from being able to squeeze in between the cars and thereby disrupting the motorcade. But, as I mentioned, there was absolutely no crowd control and the car engines started to overheat on the way to the hotel. One of the security people for the astronaut wives came back to the car that I was in and told us that they were taking our car for the wives and that we had to get out of the car. In all of that humanity, we had to look for another car which was hard to find since the crowds had gotten in between all of the cars. Essentially, the drivers were on their own to try and find their way without following the car in front of them. We finally found a place to get into and we were sitting on each other's laps and crammed into the back seat of this little car. That was probably the hairiest experience. I was down right scared. I thought what if I can't find another car to get into what am I going to do? You just don't know what you'd do.

The astronauts' car also overheated. I don't understand the mechanics but apparently when engines overheat, you can turn on the heater allowing the engine to cool so that you can go a further distance instead of stalling. Their driver turned on the heat and took a short cut through a soccer field in order to get to the hotel.

In Bombay, India, the embassy estimated the crowd to be about 1.5 million people. There was an outside ceremony. At Dacca, they estimated there to be about a million people . . .

In Kinshasa, Zaire, there was a 25-mile drive from the airport to the place that we were staying. They had us billeted in a compound of villas which President Mobutu had ordered to be built for an Organization of African States Conference and the compound was adjacent to the presidential palace and his offices. The crowd control there was totally opposite from what we found in Dacca because the policemen had these huge whips and if somebody stepped off the curb to get into the path of one of the cars in the motorcade or tried to get to the astronauts' car, they would use the whips on people.

The compound where we were staying had a private zoo which the president owned and you could hear the animals, especially at night . . .

In Tehran, Iran, the staff was invited to go to the vault where all of the Shah's family crown jewels were stored. The vault doors were on a time lock. The vault was opened—especially for us, but I believe it was still time locked. We were cautioned not to touch the glass display cases but somebody did and alarms started going off and the heavy doors started closing. The security guards and the police were doing their best to hold the doors open while we all scurried out as fast as we could go.

A lot of these events were televised also?

Most of the arrivals were televised live. In some of the cities, a national holiday had been declared on the day of the astronauts arrival.

It must have been impressive going to these various countries in the president's plane . . .

Yes, it was. One of the embassy people told me that he was at the airport for one of our evening arrivals. He said that when we approached the airport and he saw the plane, with the presidential seal and the American flag imprinted on the plane, it just gave him goose bumps. It was quite an impressive plane. When the trip was over we all gathered under the seal and had a group photograph made . . .

During the State Department briefings we were told that whenever you heard the Star Spangled Banner that no matter where you were or what you were doing, you must stop, face the flag, and put your hand over your heart. This happened in Berlin as we came down the steps of the plane and the band started playing the Star Spangled Banner. One of our embassy people said that the scene was picked up on live television.

Were you allowed to take personal items with you during the trip?

We were allowed to have one suitcase to be stored in the hold of the airplane, and a hang-up bag which could either be up front with you in the cabin or in the hold, and a briefcase. I was able to smuggle in a wig box because in those days wigs were the rage and they really came in handy when it wasn't convenient to find beauty shops. We lived out of a suitcase, a hang-up bag, and a briefcase for the entire 38-day trip.

237

The Apollo 11 astronauts and their wives receive a papal audience by Pope Paul VI in the Papal Library, St. Peters Cathedral. (NASA Photo 70-H-1576.)

One of our staff, Herb Oldenburg, had the important job to get the right luggage to the right room at the hotel when we got off the plane. He and his people would also collect the bags early in the morning on the day of departure to get them to the airport. It sort of became a game of Russian roulette whether to pack your bags and put them outside your room the night before a flight or to take your chances that you are going to get up early enough to get your bags out then. I generally put mine out at night and kept out what I was going to wear the next day and stuffed my nightclothes in my briefcase.

When we were in Sydney, I was dragging my bags out in the hallway about two o'clock in the morning. I used my raincoat as a bathrobe because I wanted to save room in my suitcase for souvenirs. I was dragging my bags out in the hall with my raincoat on, my hair was up in curlers and I was in my bare feet. There was no one there when I opened the door and looked out but after I stepped out into the hall I heard this Australian voice behind me say "Good morning, young lady." I turned around and it was the Prime Minister of Australia! [Laughter] He had been to a social function down the hall and was walking by on his way out of the hotel . . .

The trip was pretty intense. I lost 25 pounds during the trip.

In retrospect, in looking at everything that made Apollo happen, probably one of the most difficult things for the crew members was in dealing with the public. They were naturally uncomfortable around the public and now suddenly they were on front stage to the world . . .

It must have been very difficult for them. People were looking to them to say inspirational things. They were called upon to make speeches and remarks at every place they had a public appearance.

Did you encounter any strange customs that you had to adhere to during the trip?

In Australia, when we landed in Perth, the local officials insisted on coming onboard and spraying for tsetse flies. Nick Ruwe, our State Department guy, was highly offended that they would do that but they insisted on doing it anyway. We were told that when they came onboard to inspect the plane, we should just sit there and stare at them.

When the tour was finished and the astronauts arrived back in the U.S. did they give a report to the President?

Yes. There was a White House ceremony on the day we returned, November 5. During the ceremony, President Nixon officially welcomed the astronauts back. The astronauts presented the report, along with their letter, summarizing the report, to the President. The support staff and our families were also present during the ceremony.

After all of this was done, did the astronauts have an opportunity to take a real vacation?

I'm sure they did. However, about a month after they returned they were sent to Ottawa and Montreal, Canada. Apparently, these stops couldn't logistically fit into either the beginning or the end of the world trip. They were received by the Prime Minister of Canada in Ottawa and made appearances there and in Montreal.

From my own personal viewpoint, I know that when the trip was over, I was glad to get back home to my family. I had three children who were at that time 3 (Paul), 11 (Susan), and 13 (John). My ex-husband and the children met me at the White House for the welcoming ceremony. When we left to get the car from the parking garage, all of a sudden it dawned on me that there was nobody bringing me an official car right away. Reality had set in. I was just worn out. When we got home, I didn't even unpack my bags, I just sat down in a reclining chair and sat there all night. The whole time that we were traveling, the adrenaline was at its peak and I could rest from being afraid of missing the wake up call, afraid of missing the plane. All of a sudden everything is down, you don't have to wake up at a certain time the next morning. You don't have to pack a bag to put out in the hall. It was like you really came down and it took a couple of days to get yourself back to the functioning level—to return to Earth so to speak.

I'm glad that I was given the opportunity to take part in such an adventure . . . It was really quite unique and a wonderful experience and one that I never would have had had I not worked for NASA. I never would have dreamed of going to all of these places. I'm glad that I did it but I would not want to do it again.

What are some of your thoughts on the achievements made by Apollo?

I worked the protocol part of the Apollo 11 launch and it was really quite mind boggling to see. We were at Cocoa Beach a week ahead of the actual launch day and people were already gathering to witness the launch. The evening before the launch, there were people sleeping on the beaches and in their cars because there were no more hotel rooms. Some spent the night sitting in chairs in hotel lobbies. The motel and restaurant marquees at Cocoa Beach were all saying "Good Luck Apollo 11." There were a lot of well known people there for the launch. You could not help but feel that there was something big happening and you were just glad to be a part of it. My family went down for that launch. I remember when it took off, when it lifted off the pad, I could not help but think, what are they really going to experience once they get there. Are they going to get back? Because in spite of all the things that you heard about the mission being carefully planned and they knew what to expect and everything was going to go according to the flight plan, I always had a feeling in the back of my mind well, what if . . . I thought that the eyes of the world were focused on Cocoa Beach at the Kennedy Space Center. I just felt glad to be a part of it and I wanted my children to see it and to share the experience with me. My youngest child was just two-and-a-half years old and he vaguely remembers it because of the launch sequence when it starts to lift off with all of the noise and smoke. My two older ones will never forget it. I know I certainly won't. This was such a significant event for the United States and NASA and I felt privileged to have had a very small part in it.

ENDNOTES

1. The crew of the Presidential Aircraft assigned to the first half of the Giantstep Apollo 11 World Tour (from Washington to Rome) include the following: Maj. David H. Shaw, Aircraft Commander; Maj. Lester C. McClelland, Aircraft Commander; Capt. Adolph C. Zerumsky, Navigator; S/Sgt. Walter Battic, Guard; M/Sgt. James H. Brown, Steward; S/Sgt. Larry N. Coleman, Guard; S/Sgt. Errol E. Devore, Guard; M/Sgt. William W. Gibbs, Jr., Flight Engineer; T/Sgt. John R. Jester, Guard; M/Sgt. Larry L. Kerns, Flight Engineer; M/Sgt. Robert A. Koehler, Radio Operator; S/Sgt. Eugene L. Munger, Steward; S/Sgt. George R. Phillips, Guard; M/Sgt. Robert A. Rouse, Steward; M/Sgt. Darrell F. Skinner, Radio Operator; M/Sgt. Buddie L. Vise, Steward; M/Sgt. Doyle G. Whitehead, Steward. The crew assigned to the second half of the tour (from Rome to Washington) include the following: Maj. Kenneth L. Cox, Aircraft Commander; Maj. Frank O. Pusey, Aircraft Commander; Maj. Robert W. Pollard, Aircraft Commander; Maj. Donald F. McKeown, Navigator; M/Sgt. William A. Scholl, Flight Engineer; M/Sgt. Donald E. Caton, Flight Engineer.

2. In addition to the flight crew for the presidential aircraft, those that went on the world tour include the following: The crew of Apollo 11 and their wives—Neil Armstrong, Janet Armstrong, Michael Collins, Patricia Collins, Edwin Aldrin, Jr., and Joan Aldrin; NASA support personnel—Howard G. Allaway, Public Affairs Office, Office of

Manned Space Flight, NASA; Geneva B. Barnes, Secretary to the Dirctor, Public Events Division, Office of Public Affairs, NASA; Simon E. Bourgin, Science Advisor to the U.S. Information Agency (USIA), Office of Policy and Plans; Dr. William R. Carpentier, Flight Support Officer for Apollo Preventive Medicine Office, NASA; Joan Carroll, Secretary to the Assistant Chief of Protocol, Department of State; William Der Bing, Deputy Chief, Special Events Office, Manned Spacecraft Center, NASA; Robert B. Flanagan, Senior Security Specialist, NASA; Richard Friedman, Information Officer for International Affairs, NASA; Joseph Kidwell, Protocol Assistant, Office of Public Affairs, NASA; Charles G. Maguire, Staff Assistant to Deputy Assistant Secretary for Operations, Department of State; L. Nicholas Ruwe, Assistant Chief of Protocol, Department of State; Wade St. Clair, Director, Public Events Division, Office of Public Affairs, NASA; Julian Scheer, Assistant Administrator for Public Affairs, NASA; Elton Stepherson, Jr., Special Assistant to the Area Director for Near East and South Asia, USIA; William P. Taub, Visual Information Officer, Office of Public Affairs, NASA; Herbert Oldenberg, Department of State; Thoreau Willat, Voice of America; Edward S. Hicket, Voice of America; Enrique Gonzales-Reguerra, Voice of America; Petro Luis Kattah, Voice of America. There were also NASA security, public affairs people, and White House advance men who traveled ahead of the main group, leapfrogging from one city to the next, coordinating security, logistics, and any last minute surprises on the schedules. These people were as follows: O.B. Lloyd and Walter Pennino, NASA Public Affairs Officers and advance team; Frank Dukes and Arnold Garrott, NASA Security Officers; James Bertron, David Cudlip, Leonard Steuart and Edward Sullivan, White House advance team.

3. The itinerary for the "Giantstep Apollo 11" world tour from September 29 to November 5, 1969 included the following stops: Mexico City, Mexico (Sept. 29-30); Bogota, Colombia (Sept. 30-Oct. 1); Brasilia, Brazil (Oct. 1); Buenos Aires, Argentina (Oct. 1-2); Rio de Janeiro, Brazil (Oct. 2-4); Las Palmas, Canary Islands (Oct. 4-6); Madrid, Spain (Oct. 6-8); Paris, France (Oct. 8-9); Amsterdam, Holland (Oct. 9); Brussels, Belgium (Oct. 9-10); Oslo, Norway (Oct. 10-12); Cologne/Bonn and Berlin, Germany (Oct. 12-14); London, England (Oct. 14-15); Rome, Italy (Oct. 15-18); Belgrade, Yugoslavia (Oct. 18-20); Ankara, Turkey (Oct. 20-22); Kinshasa, Zaire (Oct. 22-24); Tehran, Iran (Oct. 24-26); Bombay, India (Oct. 26-27); Dacca, East Pakistan (Oct. 27-28); Bangkok, Thailand (Oct. 28-31); Perth, Australia (Oct. 31); Sydney, Australia (Oct. 31-Nov. 2); Agana, Guam (Nov. 2-3); Seoul, Korea (Nov. 3-4); Tokyo, Japan (Nov. 4-5); Elmendorf, Alaska (Nov. 5); Ottawa and Montreal, Canada (Dec. 2-3).

4. There were three major items for the astronauts to present to heads of state and other dignitaries during the Apollo 11 World Tour. These include the following:
   • Replica of the plaque left on the Moon mounted on a walnut backing. A plaque was presented to the leading official at each city.
   • Replica of Goodwill Message disc left on the Moon, eight-power magnifying glass and framed photograph. These were presented to the signers of each of the individual messages that were contained in the original message disc. Nations on the Apollo 11 World Tour that provided goodwill messages included on the message disc include the following: Argentina, President Juan Carlos Ongania; Australia, Prime Minister John Gorton; Belgium, Baudouin I, King of the Belgians; Brazil, President Arthur Da

Costa E. Silva; Colombia, President Carlos Lleras Restrepo; Congo, President J.D. Mobuto; India, Prime Minister Indira Gandhi; Iran, Mohammad Reza Pahlavi Aryamehr; Italy, President Guiseppe Sarget; Japan, Prime Minister Eisaku Sato; Korea, President Park Chung Hee; Mexico, President Gustavo Diaz Ordaz; Netherlands, Juliana R.; Norway, King Olav R.; Pakistan, A.M. Yahya Khan; Thailand, Bhumibol Adulyade, King of Thailand; Turkey, President Cevdet Sunay; United Kingdom, Elizabeth R.; Vatican, Pope Paul VI; Yugoslavia, President Josip Broz-Tito; Canada, Prime Minister Pierre Elliott Trudeau.

- Color photographs from the Apollo 11 mission (8" x 10" mounted on 11" x 14" mats; 11" x 14" mounted on 16" x 20" mats) were autographed aboard the aircraft for presentation to lesser dignitaries (i.e. ministers, ambassadors, mayors, etc.).

The family of Apollo 16 LM Pilot Charles M. Duke, Jr., are shown at the launch site in this early February 1972 photo. With Duke are his wife Dorothy and sons Thomas (4), at left, and Charles (6). (NASA Photo 108-KSC-72P-62.)

CHAPTER 11

CHARLES M. DUKE, JR.

(1 9 3 5 –   )

Air Force Test Pilot Charlie Duke became the tenth person to walk on the Moon when he was lunar module pilot of the Apollo 16 mission in April 1972. Duke and Commander John Young landed their lunar module "Orion" on the Cayley Plains near the crater Descartes six hours later than scheduled because of problems with the main rocket engine of their command module "Casper." Concern over the engine forced them to shorten their planned flight by a day, but Duke and Young nevertheless spent almost three days on the lunar surface, including nearly 21 hours outside Orion.

Apollo 16 was the second of the "J" Type missions, utilizing a modified version of the spacecraft used for the first three lunar landings. The J type spacecraft allowed crews to stay on the lunar surface for extended periods of time. The Portable Life Support Systems (PLSS) doubled the capacity of those first used by Aldrin and Armstrong during Apollo 11. Apollo 16 also had the distinction of being the second mission to feature the lunar rover, an electrically-powered vehicle that allowed the two astronauts to travel much greater distances than before on the lunar surface.

Charles Moss Duke, Jr., was born October 3, 1935, in Charlotte, North Carolina. He attended the U.S. Naval Academy, graduating in

1957 with a B.S. degree in Naval Sciences. In 1964, he earned an M.S. in aeronautics and astronautics from the Massachusetts Institute of Technology.

After pilot training, Duke spent three years with the 526th Interceptor Squadron at Ramstein, Germany. He attended the USAF Aerospace Research Pilot School at Edwards Air Force Base, California, in 1965, and was an instructor there when selected by NASA. As a pilot, Duke logged 4,200 hours of flying time, including 3,600 hours in jets.

Duke was one of 19 members of the fifth class (Group 5) of astronauts selected by NASA in April 1966. Calling themselves the "Original 19," this group of astronauts was eligible for Apollo missions, but were left out of the Gemini program rotation. Of the 17 that eventually flew, only three did not go into space on Apollo-related missions.

Duke entered the "official" mission rotation as a member of the support crew for Apollo 10. He also served as back-up lunar module pilot on Apollo 13 and 17. Duke was also Capcom on the White Team during the first lunar landing mission.

Following the Apollo program, Duke worked on Space Shuttle development for several years before resigning from NASA on January 1, 1976, and from Air Force active duty. Explaining that he worked 80 hours a week for the past few years to get to the Moon and that his work in Space Shuttle development was boring, Duke followed in the footsteps of his close friend and fellow astronaut Stuart Roosa and started a wholesale Coors beer distributorship. A self-described workaholic, Duke sold out of the business in March 1978 and went into a series of investing concerns and real estate development ventures. Duke also continued work with the Air Force as a brigadier general in the Reserves.

Duke's most profound experience after leaving NASA was a spiritual awakening as a born-again Christian. During his career as an astronaut, he and his wife Dorothy ("Dottie") had marital problems. Dottie turned to her religious faith for help and Duke followed in kind to become a Christian Lay Witness. After associating with fellow Apollo astronaut James B. Irwin's High Flight ministry, Duke started the Duke Ministry for Christ. His autobiography, Moonwalker, written with his wife, was published in 1990.

Editor's Note: The following are edited excerpts from an original interview with Charles M. Duke, Jr., conducted by Doug Ward on March 12, 1999, as part of the NASA Johnson Space Center Oral History Project.

You were one of, I think it may have been more common at that time than it is now, the Naval Academy graduates who went into the Air Force . . . How does that work?

Back in those days there wasn't an Air Force Academy. Their first class was 1959, so they allowed ("they" being the Defense Department) 25 percent of West Point and Annapolis graduates to volunteer for the Air Force. So the Air Force was basically culling out their regular Officer Corps from West Point and Annapolis . . . So before we graduated, we said, "Well, we'd volunteer for the Air Force." And I'd fallen in love with airplanes at the Naval Academy rather than ships; and I knew that's what I wanted to do. The airplanes I thought were better. You could stay in the cockpit longer in those days.

How did you get hooked on airplanes?

My first recollection of flight was back in the early '50s. I was with a friend, and we'd just gotten our driver's license and were driving along. He had a little old convertible, and I looked up and there was a contrail going over. In the early days of jets, you didn't see many contrails back then. And I said, "Gosh, it'd be nice to make a contrail. I wonder what that'd be like?" And I started dreaming about flying airplanes then. I went on to the Naval Academy and they gave me a couple of rides in an open-cockpit, bi-wing seaplane called the N-3N Yellow Peril. I was hooked from that moment on.

Didn't get sick?

No. I got seasick, but I never got airsick. [Laughter] And maybe that's another reason that I decided to go [in the Air Force], because I really did get seasick. But I never did get airsick. [Laughter]

Getting back to the astronaut selection process and after you'd applied, you started moving through the process, the physicals and all of that. You get down to the final interview . . . do you remember who participated in that?

Yes, I do. John Young, who I ended up on the Moon with, and Mike Collins, Deke Slayton, and Warren North were the four that I remember that were on that committee. I don't remember the nature of the interview. It was more a get-to-know-you kind of "What's your motivation?" interview because they had our backgrounds . . .

And we had gone through the physicals and nobody had any problem with the physical. And, you know, we'd taken some preliminary tests but with a master's degree and all I knew we were in the running . . . I was really getting pumped up by then. I really wanted to get picked.

Did you have the impression, at that point, the extent to which public relations would be a prerequisite for the job?

No, I didn't . . . I didn't perceive that would be part of my job. I perceived that they were certainly in the limelight—the original guys and the astronauts who'd been selected. Gemini was going on and I knew that they were in the limelight but I never realized that this was a big part of the job . . . It turned out I ended up liking that but, I didn't realize that was going to part of it when I started.

So they really weren't as up front as they might've been?

No. It was something that I'm not sure anybody would've turned it down because they had to go out and make a speech. But, you know, it just turned out some people are more comfortable doing that than others. And I love to meet people . . . when your heart's in something, you can really talk about it and be sincere.

Getting back to [Deke] Slayton and [Alan] Shepard. How did you perceive, at that time, the relationship between the two of them?

I sense it was a close relationship. Of course, I was in awe when we got here, you know. Gosh, here's Alan Shepard, and Deke Slayton, you know, and Wally Schirra and all those famous astronauts who I just admired and looked up to for years. I sense that Shepard, Deke, and that whole group were real close; and their working relationship, which I didn't quite understand in those days, was very tight. I didn't see any competition.

Do you think that whatever criteria Deke and Al used in selecting crews was fair and effective and did you ever sort out what that criteria was?

I never sorted that out. [Laughs] I've been asked that many, many times. "How did you get picked?" I said, "I don't know." Even to this day, I'm not sure how the crews were selected. I got an inkling that Deke and Al sat down and said, "Okay, who's going to be the next commander?" And, "Okay, it's John Young." And John Young gets called in. And then Deke says to John, "Well, here are the guys that we think we ought to fly, what do you think?" So in some way, through that process, you got selected and then it went up to the directorate, from there to Washington . . . it is a mystery to me. There was no "check out the squares," you know, "fill this board of squares," "do this task," "do that run," "do this thing." And you get all these squares filled and you're going to get a flight. We didn't have any of that. It was just, "do your job."

When we first got here we started training. Everybody did geology. Then we did spacecraft systems for four or five months. And then everybody got assigned to some sort of little engineering oversight job. I remember Stu Roosa and I got assigned to Frank Borman, who was head of the Propulsion System for the side of the Astronaut Office. These things were all sort of unofficial, I think, as far as organization within the Office but that was where you concentrated your effort.

And so, on behalf of the Astronaut Office, you went to design reviews on the Saturn or the engines or the guidance system which took us to Marshall and to Kennedy. I remember we went to Ames a couple of times to try to "fly the Saturn V in orbit manually." Ames had a program and some sort of simulator that Stu [Roosa] and I went out there and we could fly the Saturn V into orbit from the pad. There was a program in the guidance system in the Saturn in the Instrument Unit [IU] and it was connected through software to the controller in the command module. The commander could sit there and, with the eight ball, fly this thing into orbit like you do an instrument landing in an airplane. We showed that we could do that . . . We never had that capability, at least not in first stage but in second and third stages we had that ability. It was incorporated into the software.

What was your assignment on Apollo 12?

I wasn't really assigned to that flight. I was backup on 13 so I was not involved officially in Apollo 12. I had been support crew for Apollo 10 and helped develop the lunar module procedures during that time for activation and checkout. I was a Capcom and support crew during the training. Then for the flight I was Capcom for the lunar orbit activation, checkout, and rendezvous on Apollo 10. As a result of that, I probably had the most experience of any of the astronauts in the lunar module that was not on a crew. And so I ended up on Capcom and doing the same job on Apollo 11.

Did Armstrong have a vote in that?

Apparently he did. Neil had come by and said, "Charlie . . ." (I believe it was Neil and Buzz . . . maybe just Neil) " . . .we'd like you to do activation and checkout for us for the landing." And I said, "That'd be a great honor. I'd love to do it." And of course, that's what I did.

That really wasn't the normal rotation. You wouldn't normally go as a Capcom for one flight right into the next, would you?

No. Normally, you worked support crew, then you went into maybe a backup crew and then on into a flight crew. So about every three flights, you had some progression and that's what happened for me. From [Apollo] 10 to 13 to 16, was the prime way it went for me;

and it just turned out that I didn't get on the backup crew until after 11. I think they didn't announce that crew until after 11, and that's how I ended up Capcom for Neil and Buzz for the landing.

Prior to that, I had taken over the lunar module engine oversight. The descent stage engine was doing really well, but the ascent engine was in trouble. This was in early 1968. They were trying to qualify it and were having some instability in the engine. And that's a critical engine. If it didn't work, you were going to either crash on the Moon or be trapped on the Moon.

It was kind of like the service module engine in the command module. It just had to work.

It just had to work . . . and so, George Low, who was Apollo Program Manager at the time, organized this committee to decide what we were going to do. Would we get this thing qualified or should we have a parallel development?

We met for a couple of months and visited with various contractors. We listened to proposals from Bell [Aerosystems Company in Buffalo, New York] about how they were going to fix theirs, and we went to Rocketdyne and they said, "Well, we could do it." And it ended up, about six months later, we decided to have this big meeting with [George] Low. There was a vote. "What are we going to do?" I remember everybody sort of looked at me. "What does the crew want to do?" And the response was, "Well, I think we ought to go with a new contractor." We'd been parallel this time; Rocketdyne was selected and within a month they had their engine qualified. It turned out it was a great engine for . . .

You changed horses in midstream. Went from Bell to Rocketdyne . . .

Which was rare that close to a launch. It was beginning to impact the schedule because Kennedy said, "Land by the end of the decade," and now we're at late '68. It was a critical time, and that was a big decision because to change a contractor that close . . . We did it and I thought it was a bold move by Low.

251

The crew of Apollo 16 pose during a training exercise at the Kennedy Space Center. Shown left to right are Lunar Module Pilot Charles M. Duke, Commander John W. Young, and Command Module Pilot Thomas K. Mattingly II. (NASA Photo 72-H-249.)

Has the thought crossed your mind that you may be remembered as much in history for your role on Apollo 11 as for your own lunar landing?

You know, I was just pleased to have been part of that team. I loved the Mission Control team. I thought we had the greatest bunch of guys and gals in that Mission Control. They were dedicated, young, but experienced; and it was great working with them. All the flight directors that I worked with, we all hit it off really well. And [they were] very confident, very cool, in all of the problems that we experienced.

Now that's interesting, because it really wasn't always that way. If you go back and talk to some of the early groups, there were some battles between the astronaut representatives who were sent out to the remote sites, between Deke's [Slayton] guys and

Chris's [Kraft] guys, as to who was going to be in charge. And it's pretty clear that, through Mercury and Gemini, all that got sorted out; it really was a nicely functioning team by the time you guys got into Apollo.

Yeah, I'm sure it was, and I'd heard some of those stories but I never experienced that. By the time I got there, Apollo 10, I had done one of the trench jobs. I remember in Gemini, there was an astronaut and also a Capcom that was an astronaut in the booster position. And I did that in Gemini XI and XII. I really didn't know what I was doing there but even then, I felt smooth-running teamwork had evolved into the Mission Control, and there wasn't any of that "I'm in charge here" dynamics.

It certainly gave you a lot more confidence when you had to rely on the judgments of the ground and you were the one in flight.

Which we did. On Apollo 16, an hour before we were to land on the Moon, we were on the backside of the Moon and Ken [Mattingly], in the command module, had to burn the SPS [Service Module Propulsion System] to boost his orbit up . . .

And this was after the lunar module had separated?

We were separated. We were a mile or two apart and we were within an hour of landing. The next half revolution around, we were going to start our descent. And on the backside, he was to boost up to a 60-mile circular orbit so that we could have the right phasing for the rendezvous if we had to abort. Well, he reports a real problem in the engine. When he powered up the secondary control system, it wouldn't stabilize the engine and it was wiggling back there. And he thought that the thing was going to shake him apart. So John says, "Don't burn."

If your heart can sink to the bottom of your boots in zero gravity ours did, because there we were, two years of training, 240,000 miles away, an hour before the landing on an orbit when you can look down at your landing site eight miles beneath you, and they're about to tell you to come home. And that's what we thought was going to happen, because it was, according to mission rules, an abort. So we came around, reported no circularization burn and they said, "Roger, stand by. Start getting back together." We started a slow rendezvous, thinking that was it, but they said, "We'll look at it."

I don't know all the dynamics which went on in Mission Control, but four hours later, after a couple of more revs, they kept briefing us and said, "Well, we think we're isolating the problem and think we can work around it." Later on, I saw a video that they'd taken at Mission Control, and I can remember Chris Kraft sitting there, scratching his head saying, "Let them have a go at it." And he said, "Go," and so Jim Irwin said, "We're go for this attempt," the second time around. We didn't know what was wrong but they figured it out. At that point in our flight to have management say, "Yes," to a landing I thought was really tremendous because it would have been a lot easier to say, "Well, come on home, you guys," "We don't want to risk it." You know, "We've done it before, and . . ." But to let us go ahead and land was really terrific. Almost every flight, you could see that the Mission Control team had a great hand in aiding in the rescue or continuation of the mission or overcoming some problem that we didn't have a clue of what it was in the cockpit.

There were a couple of problems on the Apollo 11 descent that, as I recall, were pretty unnerving. One was the continual communications dropouts and telemetry dropouts with the lunar module. Of course the other was the computer alarms. At the time, do you remember which of those or other problems were uppermost in your mind?

Communication dropouts were a nuisance more than a danger, but a computer problem was a showstopper. I vaguely remember reading something about a 1201 alarm. I didn't really know the consequences of those alarms and you didn't have time to break out your guidance and navigation [G&N] checklist to go to the emergency procedures and find out what that was. But fortunately, Steve [Bales] and those guys on G&N console knew.[1]

Then later on, the most critical was, of course, the fuel state. We didn't quite have the tracking right in those days. We didn't understand what the gravity anomalies [mascons][2] were on the Moon, so we were a little off in position. And when they pitched over to look at the lunar surface, they didn't recognize anything and they were going into this big boulder field. Neil was flying a trajectory that we'd never flown in the simulator or in our integrated SIMs with Mission Control; it was something we'd never seen. We kept trying to figure out, "What's going on? He's just whizzing across the surface at about 400 feet," and all of a sudden the thing rears back and he slows it down and then comes down. And I'm sitting there, sweating out . . .

You could see all that from the telemetry . . .

Yeah, we could see all of that. I had that plot on my screen. We were getting critical fuel-wise. I remember I was giving them this running commentary. We were down to the last couple of minutes and Deke Slayton is sitting next to me. We were glued to that screen and I'm just talking and talking and telling them all this stuff. And Deke, I remember, punches me in the side and says, "Charlie, shut up and let them land." [laughs] "Yes, sir, boss." So I got real quiet, and the tension began to rise in Mission Control. We had a 60-second call and . . .

That was 60 seconds to abort?

That's right. When Mission Control said, "60 seconds," it meant you had 60 seconds to get on the ground. And the problem was fuel. We wanted enough fuel remaining in the descent engine that when he throttled up, he would get a positive rate of climb and start up before we had an abort stage because that was critical . . . And then the next call was "30 seconds." And so I called, "60," and they were still in the air. And I called, "30" and "Man, it's getting close." And then, of course, the dust was flying. And then I heard, "Contact. Engine stop," and I knew we were on the ground.

This was after the 30-second call?

After the 30 seconds, right. Later on you look at the data and there was between 7 and 17 seconds of fuel remaining.

That was 17 seconds remaining before you would have had to make an abort?

Before I would've had to abort . . . Now whether Neil would actually have aborted or not, I don't know. Had I been the commander and I was within the dust and if somebody even called "Abort," you know, and you were 10 feet off the ground, what were you going to do? Well, I'd probably have landed.

Anyway, we landed before the 30 seconds were up and, of course, everybody erupted in Mission Control. Then we herad the [Neil Armstrong] famous lines about, "Houston . . . Tranquility Base here. The Eagle has landed." And so we made it and it was really a great release . . . Then Gene Kranz got us all back to work because, at that point, we had to make sure this thing was safe, that we didn't leak anything and if we had to, we could lift off again quickly in an emergency.

Your comment at that point was obviously very spontaneous.

It was. I was so excited. I couldn't get out "Tranquility Base." It came out sort of like "Twangquility." And so it was, "Roger, Houston. Twangquility Base here." Let's see, what did I say? No, it was, "Roger, Twangquility Base. We copy you, down. We've got a bunch of guys about to turn blue, we're breathing again. Thanks a lot." And I believe that was a true statement. It was spontaneous, but it was true. I was holding my breath, you know, because we were close.

Getting back to the simulators. One of the things that I've been told is that almost invariably the commanders in landing on the Moon and in landing the Lunar Landing Training Vehicle (LLTV) tended to use a lot more propellant and take it down a lot closer

to the wire than they did when they were in the lunar module simulator. I don't know if that was your experience . . .

Well, I didn't get to fly the LLTV out here at Ellington . . . After
Neil Armstrong ejected from the one that went out of control, they
only let the commanders fly that thing. So, I didn't get to fly it.[3]

**I think that was generally viewed as probably one of the most dan-
gerous parts of training.**

Yes, it was. As I recall, the LLTV had a jet engine that took out
five-sixths of the Earth's gravity and then used lunar module-like
thrusters to give you the rest of the one-sixth gravity. We had
something that was similar to that up at Langley, but it was on a
wire and you slid down this wire at sort of one-sixth type gravity. I did fly that, but it wasn't very good.[4] I thought the
simulator was a lot better. The only thing with the simulator was
that it was bolted to the floor so it didn't give you the motion
cues that you had in actual flight. But other than that, the simulation was really, really good. And I believe you're right. In the
simulations, you tended to be a little bit more cautious than in
the real world.

For some reason, when we came down to land, John [Young] just
continued right straight on down and we were basically in a constant descent of some rate throughout our landing. And I remember
at about 25 feet, he did level off, and we just sort of stopped. And I
remember saying, "Okay, John, you've leveled off. Let her on down."
I was feeding him the information that he needed for the velocity
and the control and the rate of descent. After a second or two of
hovering, we began to sink down about one or two feet a second,
and touched down with plenty of fuel left.

Of course, that's a function of experience. We were the fifth
landing and while you'd never done it yourself before, everybody
else had done it, and you just gleaned that experience from the
briefings and the debriefings and talking to guys about it. And so
we felt real confident, when we were there, that we were going to
have plenty of gas for the trip.

Of course by the second mission [Apollo 12], Pete Conrad was able to set down very close to the planned landing site. You must have surveyed it.

Yes. We'd figured out the gravity anomaly deal and so we were able to track and tell the computer accurately where it was and where it wanted to go. That guidance system would take you right down if you gave it the right information. It was a good system.

The mascons, as I recall, were present because of the fact that the Moon isn't made of one consistent mass. It has mass concentrations like raisins in a cookie that affect the gravitational pull on a vehicle.

That's right. As you go over one of these, it pulls you down. More gravity, and you sort of start going in an orbit like this and as you go down, it slows you down and so speeds you up. We ended up not being where we thought we were as we projected two to three revs later. So we had to figure that out . . .

One of the things that struck me, I don't know if you remember it as clearly, is the reaction at Mission Control during the simulations for the lunar landing, when they finally got all of the final guidance software in from MIT and had the actual program that was going to be used for landing on the Moon . . . On your mission, of course on all the lunar missions, one of the things in the lunar module simulator that people today would probably find a little incredulous is that you didn't use computers to generate the television views out the window. You had a big board adjacent to the simulator where the lunar surface itself was actually mocked up in all the detail that the scientists could give.

That's correct. Our landing site was selected by the site selection committee of scientists in various disciplines; and it had been selected from photographs taken on Apollo 14. Stu Roosa, as he orbited, had a mapping camera and their orbit took them over the Descartes highlands region. And so, it was decided that we would go land there. They took some of his photographs and made a

Astronaut John Young, commander of the Apollo 16 lunar landing mission, stands at the Apollo Lunar Surface Experiments Package (ALSEP) deployment site during the first EVA at the Descartes landing site. The components of the ALSEP are in the background. The lunar surface drill is just behind and to the right of Young. The drill's rack and bore stems are to the left. The three-sensor Lunar Surface Magnetometer is beyond the rack. The dark object in the right background is the Radioisotope Thermoelectric Generator (RTG). Between the RTG and the drill is the Heat Flow Experiment. A part of the Central Station is at the right center edge of the picture. This photo was taken by Charlie Duke. (NASA Photo As16-114-18388.)

---

mockup of our landing site. It turned out that the photographs had a resolution of about fifteen meters. In other words, anything less than forty-five feet in diameter, you couldn't see in the photograph. But the major features, you could see. And so they built that into the model and they put this model on a big board. There was a TV camera that ran on a track above it so that was the view you had in your window . . .

And you were simulating a landing?

259

Yes. Thousands of times John [Young] and I came in for landing and we'd pitch over and recognize features. One crater we called "Gator," and another one was called "Lone Star." I could look out the right window of the simulator, my right side, to the north and see North Ray Crater up there and John could look out his side and see Stone Mountain out to the left. These were names that we had given these prominent features in our landing site. When we really did it for the first time, I mean for real, in flight, as you recall, the lunar module trajectory was such that the first seven or eight minutes of the descent was with the window pointed out at space and you couldn't see the lunar surface.

So you kept the engine in front of you, slowing down . . .

Yes. So you were slowing down. Now you could've rolled over 180 degrees to put the windows down, but then you had a problem with communications. So we chose to land, or start down, with the windows pointed out to space and just depend on the LM to bring us in. At 7,000 feet, the guidance program maneuvered the vehicle to windows forward down, and for the very first time, at 7,000 feet, you saw the lunar surface. Well, I mean it looked exactly like the mockup. There we were. "John, there it is!" you know, "There's Gator. There's Lone Star." We'd had some debate about getting up to North Ray Crater during the training because, in the photographs, it looked really rough. I looked out the window and looked north, and said, "John, I can see North Ray. It's smooth up there. We're going to be able to make it." About that time, I'm just out the window, and John says, "Give me some information, Charlie." And so I get back in and start helping him land because we've got to pick out a landing spot.

About three months before the mission, I had this dream about John and me driving the rover up to the North Ray Crater and we came over one of the little ridges, and there's a set of tracks in front of us. It's rover tracks! Well, gosh, you know, this was all in the dream and we reported to Mission Control. We started following these tracks. Well that dream was so real that when I got to the Moon I wanted to look north to see if I could see that set of tracks

... Well of course there wasn't any set of tracks. I did figure out, as I looked north, that the surface wasn't as rough as we expected. So we ended up maneuvering, and we were like 300 feet above the surface, and John was fully in manual.

The lunar module had dual controls like an airplane. The commander was on the left side, like the captain of the airplane. I was on the right side, like the co-pilot. And I had my throttle and control stick.

Of course you're not seated. You're standing.

We're standing. We were anchored to the floor by a set of cables. As you stood there, in front of your position, there was a window here. I had a little abort guidance computer with some other switches. The main instrument panel was in between us in the center and the main computer was in this point. I'm standing and we have these cables that pull up, they're bungee-type things and you could hook on and anchor yourself on the floor. This was necessary because, if something went wrong and the thing started rolling rapidly, you wanted to be anchored . . .

I guess you couldn't afford the weight of seats.

No. There was no room, really, for seats. You really didn't need any in weightlessness or in one-sixth gravity. So anyway, we had trained that John would land and I would provide him all this information to help him down to land it. And if we had any emergencies during the final stages of descent, I would handle that because I could reach over behind him and pull circuit breakers. I could reach all the switches that were necessary to overcome any emergency. That was the way we had trained and if he had a problem with his control stick, or throttle, I'd take over and he'd perform the secondary role.

Well, it turned out everything worked right, and so I fed him the information he needed to make this landing and kept everything running right. And so he did the actual landing, and it was a great job . . . A great Navy landing! We hit solid and stable. We'd picked out a great spot. The lunar module could land on a ten-degree slope; but if you did, it was tough because the experiments were around at the back of the lunar module and that meant they would have been above my reach because I was standing downhill trying to reach up. We couldn't have done some of the experiments. But it turned out we were within one degree of level, and so we were able to work around the lunar module.

Did you kick up the amount of dust that Neil and Buzz did on the first landing?

It's hard to say. We did have a lot of dust. In a comparative sense, I would imagine it was about the same, even though our landing site was considerably higher in elevation than theirs and in a different textural context of the Moon, as far as the geology went. The dust was probably the same. I remember we almost had the surface obscured at about 20 feet. When we leveled off at 20 feet, I remember looking out and you really couldn't see through the dust that was being blasted away.

We had selected what we thought was a good landing spot. No major craters, and so we landed. It turned out, though, that when we got out the next day for our first EVA [Extra-Vehicular Activity] and I went around to retrieve the Apollo Lunar Surface Experiments package, which was called the ALSEP, there was a big crater about two meters behind us that we hadn't even seen.

Hadn't seen?

And if we'd have landed like three meters back to the east, we'd have had the back leg [of the LM] in that crater.

Was it deep enough to have tipped you [the LM] over?

Probably not. If you'd gone back another six meters, I mean, this was a pretty big crater and it tendered out maybe 15 feet deep. It would have been hard to work because I'd have been standing downhill in the crater trying to get this ALSEP out. It was amazing how things like that were sort of camouflaged. Without the right lighting conditions, you could miss some of these subtle features.

The Moon's surface had some unusual reflectivity characteristics that I guess accounted for why you wanted to land with the Sun relatively low on the horizon.

The Sun low on the horizon gave us long shadows and generally that was very helpful. You realized if you were landing on a slope that was very bright, it meant it was tilted towards you. If it was very dark, it was tilted away from you because you were in the shadows getting into the shadow side. So we tried to pick a spot that was sort of an average brightness . . . without any major rocks, boulders there. Also, you use the shadow of the lunar module to judge altitude. For instance, if you lost the landing radar at the last 200 feet, as you got closer, the shadow came in and you could use the shadow to give you some sense of altitude. And so it was very important that we land with a very low Sun angle not only because of the temperature of the lunar surface, but also for the landing aids that we needed.

One of the things we haven't touched on in detail before we continue on with Apollo 16 (your mission) is your role on Apollo 13. One of my favorite political cartoons from that era (I think it was a Bill Mauldin cartoon) showed three very glum Apollo 13 astronauts sitting in their suits, getting ready for launch, with their helmets off, covered with the measles . . . And one of them looked at the others and said, "Well, at least none of us is pregnant." . . . And of course you had a pretty direct role in that episode.

Oh the infamous measles.

I'd like to get your recollections of how all that came to pass.

I was the lunar module backup to Fred Haise on Apollo 13. John Young and Jack Swigert and I were the backup crew. In those days, you had two crews [prime crew and backup crew] for each mission and you trained in parallel so that the backup crew could take your place if something happened to the prime crew. The thought was that, you know, they [the prime crew] might have an accident or they could get sick or something like that, and then you'd have a replacement for them. You wouldn't have to abort the mission.

I guess about two to three weeks before flight, our son Tom was three, and he had a little friend named Paul Hause who was the son of some good friends of ours down in Houston. We were off for the weekend with the Hauses and sure enough, we came back a week later and Suzanne Hause called and said, "Paul has got the measles." I said, "Oh Lord" and I caught the measles from Paul, this little three-year old. And so I'm down there training all this time and I break out with the measles and go to the doctor because I'm pretty sick. And they get all excited, of course. (I forgot who the flight surgeon was down there.) But anyway, he gets all excited and starts testing everybody a couple of weeks before the flight. It turned out that everybody had had the measles except for Mattingly. So Lovell and Haise were immune, but Mattingly wasn't. So there was this big debate: "What are we going to do?" Finally the decision was made, "Take Mattingly off. Put in Jack Swigert." And they could launch if they thought they were able to do that.

This was only about a week before flight?

A week before launch. So I guess they had maybe two or three days of training. The movie Apollo 13 seemed to imply that Swigert wasn't ready, and that he was sort of a fill-in and really wasn't qualified. That wasn't true. Jack was a real good command module pilot. We were ready to go as a crew and it showed the beauty of the synergy of all of our training, that you could take somebody, a week before liftoff, stick him in, and everybody felt comfortable . . .

Lovell seemed to think that they were ready to go, and so they launched. By this time I think I'd gone back home. I was back in Houston (I think) for the launch, and when the explosion occurred, I was home in bed. John [Young] called and said, "Hey, they had this explosion and [there's] a real problem. Come into Mission Control." So, Ken and I, and John, showed up at Mission Control with some of the other guys, and that started, if I recall, 35 hours of work either in Mission Control or the simulator—as John, myself and others were figuring out the procedures to power up the lunar module, to get them back on a free return trajectory and recover them.

Just to set the stage a little for that, the spacecraft-combined lunar module/command module were about two-thirds of the way to the Moon when that oxygen tank exploded.

It was 55 hours out.

And so, that disabled the command module. I know that one of the things that people were greatly concerned about at that time was that procedures and step-by-step checklists that had been worked out months and months in advance now suddenly were out the window.

Exactly.

And you guys had to then figure out, "All right, how do you run this new spacecraft arrangement to keep from getting in any more trouble?"

We had to not only figure out how to power up but to get them back on trajectory. We practiced and developed those procedures in the simulator. We felt like we had a good handle on it. But then the problem came: "How are you going to make this thing last for 99 hours?"

The lunar module . . . which was designed for three days?

It was designed for three days. It was designed for two people, not three people. And so we had electrical power; we had oxygen concerns; we had water. All the consumables that were necessary for life had to be shepherded, if you will, very carefully. And to be honest, for the first 25 hours, I didn't think we were going to make it. I thought that something was going to run out. But by the time they did the burn to put them back on free return, which was, if I recall, something in the 70-hour timeframe, when we whipped them around the Moon and started back, it started looking better and better to me. And my thoughts changed to, "If we don't screw this up, either in Mission Control or onboard, we've got it made." And sure enough, everybody did a great job; and, I mean, the miraculous things that Dick Johnston's guys did to get the lithium hydroxide working . . . Thank God for gray tape. You know, every flight had two rolls of gray tape; and then the electrical guys figured out how to take power from the lunar module and go back into the command module and keep those batteries charged . . .

That's right. Because regardless of whether the lunar module got you back, you had to have the command module to reenter.

That's correct. And so, you know, it was a tense time during the whole procedure . . . 99 hours of drama, or thereabouts, till they separated and reentered . . . I remember we had figured out, in the simulator, that they had a series of maneuvers to do right before reentry because we had never separated this whole stack of the command module, service module, lunar module combined for reentry. It had never been designed for that.

They were going 25,000 miles an hour at that point.

Right. And accelerating . . . we were concerned about how they are going to be reentering. Could we crash them together? We had to figure out what was the best attitude and we'd done that, but it required a number of maneuvers to get it in the right position. And the more we thought about that, the more concerned I became, because we could still be maneuvering and reenter and not get it all done, or we get to gimbal lock if we have a problem with a jet. And so it was a real moment of decision, if you will. As I recall it, John and I went to (I think it was Gene Kranz) and said "Gene," you know, "why don't we just take what we got and just separate and let's just go?" And we did one or two little maneuvers, but we cut out some of them and that's what we agreed to do. And sure

enough, everything came back in; and we didn't have any problems at all with collisions.

Did Mattingly ever thank you?

When I caught the measles and he was off the mission, he was
really, I think, sad . . . and especially after the explosion. You know,
he had that sense of duty and that's where he should be . . . But
after the recovery it was announced that we would go on into
Apollo 16 together.

I don't ever remember us talking about it. It was never a
moment of, "Charlie, how could you do that?" . . . you know, it was
just one of those things that happened. By the way, after the
measles, it turned out on Apollo 16, Fred Haise was the backup
commander for us. We were climbing into the command module
on the launch pad and Guenter Wendt and the team were up
there. John gets in, and I'm the next in on the right side. As I start
to climb in, I reach in and look over and taped to the back of my
seat was a big tag that said, "Typhoid Mary." [laughs]

**Getting back to Apollo 16, as you and Young prepared for your
landing what kind of advice did you get from the prior crews, from
Apollo 11 through 15?**

On every mission after it was over, we had a day of debriefing that
was basically just the Astronaut Office. Of course 11, 12, and 14
were in quarantine after the missions, and the whole Astronaut
Office went over and spent a day just talking about procedures,
attitudes, and feelings, all of the things that you want to know
about after the flight's over. You sort of get more of a feeling of
what it's like, rather than just the technical procedures of it.

A lot of suggestions came out of those debriefings. For instance,
I remember that Apollo 15 was to be the first flight of the lunar
rover, and John and I were scheduled to be the first crew with the
rover. But they canceled Apollo 18 and 19 and moved what was the
called the "J" missions from 16, 17, 18 up to 15, 16, and 17. So
instead of being the first flight with the rover, we were the second.

John and I had monitored the development of the rover and one of the concerns we had about the rover was the seatbelt. The seatbelt was very difficult to buckle and get cinched in on the lunar surface because in the suit, you can't see down into the part where it connects on a special bar down on the side of the rover. It was sort of a blind connection. It turned out, they had a real difficult time on Apollo 15. When they got back that's what we discussed; and as a result of that, we went into redesign for Apollo 16. So John and I didn't have any trouble. We just reached over and hooked it in and then flipped it over and locked ourselves into the seat. So the debriefings really were important, I thought . . .

The lunar rover was an incredible machine . . . it revolutionized lunar surface exploration. Instead of 400 yards, you could go four miles in any direction. Our objective, of course, was the Descartes highlands of the Moon. It was a valley eight to ten miles across, and the objective was to explore the south to a place we called Stone Mountain and then to the north, three or four miles, to a place called North Ray Crater, which was at the base of the Smoky Mountains, (these were names that we had selected). With a rover, you could do that.

You know, we took 40–50 minutes to drive south. We had trained, so I was the navigator; and John was the driver of the rover. The TV camera couldn't be on during our drive across the Moon (the antenna had to be pointed right at the Earth to get a TV picture), and so as we drove the antenna whipped around and would never stay pointed. It wasn't gyro-stabilized, so we never had TV back in Mission Control while we were under way. To cover that gap, which might be as much as two kilometers, I was taking pictures and describing the terrain we were going over. I was sort of the travel guide for Mission Control—the eyes of Mission Control during that time. I had a set of maps that would take us from lunar module to Point A, Stop One. And these maps were the same photographs that had been taken on Apollo 14 of our landing area, and so it was like you were looking down. Unfortunately, once you get on the surface, some of the features just disappear, it's not like looking down from altitude. You could see the major features, like Stone Mountain, but if you were looking for a small spot like Plum Crater,

Astronaut John Young, shown with a sample bag in his left hand, moves toward the bottom part of the gnomon (center) while collecting samples at the North Ray Crater geological site. Note how soiled Young's Extravehicular Mobility Unit (EMU) is during this third and final Apollo 16 EVA. The lunar rover is parked at upper left. (NASA Photo AS16-117-18825.)

which was 1.7 kilometers to the west of us, it was like you couldn't see the forest through the trees type deal. You were just too close. But the maps were really good. We landed within a couple of hundred meters of where we thought we were going to land . . .

The lunar rover had a little directional gyro. There was no magnetic field on the Moon, so a magnetic compass wouldn't work. We had a little gyroscope that was mounted in the instrument panel of the rover, and we pointed it down-Sun and it was the old Navy lubber's line: You had a bar come down across it, cast a shadow on the gyroscope compass card, and we assumed that shadow was west so we just turned the card until 270 was up underneath that shadow; and that was our direction. We had a little odometer on the wheel that counted out in kilometers and that was our distance.

Generally, our traverses were egg shaped, elliptical maneuvers. We'd start out in one direction and we'd make a big loop and come back to the lunar module six to seven hours later. That was the plan. You never really worried about getting lost up there because everywhere you drove, you left tracks. If you really were unsure of your position, it was easy just to turn around and follow your tracks back.

Those tracks are probably still there?

I'm convinced they are, unless there was a meteorite impact nearby that, you know, created a big explosion.

The car was amazing. It was electric, four-wheel drive, and it would climb a 25-degree slope. Going up Stone Mountain, it felt like we were going out the back of the seat, because it was a pretty steep hill.

We got up to our objective, which was a place called Cinco Craters, and we turned around and sort of started back downhill and, golly! Then you really saw how steep it was, because it felt like you were going to fall out the front of the rover. Fortunately, we found a little bench, level area and we parked the car, and then we started our experiments. That was probably the most spectacular view that we had on the lunar surface.

We were three-quarters of a mile to the south and several hundred feet above the valley floor. From this advantage, you could look out all the way across the valley that we had landed in. You could see, in the distance, Stone Mountain and North Ray Crater. And there right out in the middle was our little lunar module. Looking off to the northwest as far as the eye could see, was just a rolling terrain of lunar surface, shades of gray. It was really an impressive sight. My only regret of the whole mission was that we didn't take enough pictures with people in them.

What scientifically were you looking for at Descartes?

The major objective, of course, was the geology and the photo-geology interpretation of our landing site. There were two major volcanic-type rocks: a very viscous rock that bulged up and caused the Stone Mountain topographical relief, and then down in the

Cayley Plain, the valley was another kind of less viscous rock that flowed out. We were looking at a contact between those two geologic features to see if there was any. It turned out that our landing site produced very little volcanic rock.

The major rock was breccias and igneous rock, so we had very little volcanic material. When we started describing this, I suspect that our geology team back in Mission Control were thinking "We wasted our time on these guys. They're not looking at what they're doing." But as we did more and more, they realized that this really was a unique landing site and was not like the Mauro, that it was really different. And so the rocks we collected were a unique suite of lunar materials.

The other objectives, of course involved the Apollo science package, which included a heat flow experiment which was to measure the heat coming out of the Moon. It included a magnetometer, which was to measure any residual magnetism of the Moon. It included a spectrometer, which was to measure the gases escaping from the lunar surface and two seismic experiments: one active, one passive. That was the basic science package.

Once we got the rover off the lunar module and put the TV on, I pulled out the science package. I fueled it with the RTG, which was a radioactive thermal generator (a little plutonium source). I put that into the cast that would generate the electrical power for the experiments. I hooked the packages onto the edge of a bar and remember throwing it up in the air and hooking it in my elbows. I started jogging out to the deployment site, which was a couple of hundred meters to the west of where we landed.

On the way out there (I'm jogging out), one of these packages falls off the bar and bounces across the Moon. "Oh my Lord," was my thought. I'd blown the whole deal and broken all the experiments. Well, it turned out that the package was pretty robust, and so I hooked it back up. I looked around real quick to make sure that nobody had seen that but unfortunately the TV camera was pointed right at me, and so everybody had seen this. But I recovered, and we went on out and deployed everything.

I was drilling some holes into the Moon for the heat-flow experiment while John was putting up the central station and the data area. It was a "spaghetti bowl full of cables" around this thing, with all of the experiments attached. We said, "You know up on the Moon with one-sixth gravity, these things are going to coil up like spaghetti." And sure enough, that's what happened. Unfortunately, John got one [of the cables] wrapped around his foot. He ran off and pulled a cable loose. That [cable] was the data source collector and power source for the experiment. We lost the heat flow experiment, which was tragic because I had worked hard on it, and the principal investigator was a real great guy, and, you know, we wanted to do a good job. That was the only real major problem we had as far as experiments go. Everything else worked right.

House Rock was one of the features that you guys encountered. That was a very spectacular sight on television because it loomed on the horizon, it looked so big. What was your first sensation when you saw that?

My first sensation was that it wasn't very far away. And John's sensation was, "That's a big rock!" I said, "Oh no, John. Come on. It's just right out there, let's go down there." Well, there's a problem on the Moon. Your depth perception is thrown off because you're looking at objects you've never seen before, so a big object far away looks very similar to a smaller object close in. You don't have any telephone poles or houses or trees or cars to sit and judge scale like we have down here on Earth. In my mind this rock was sort of average size and was just out there and "Let's go do it." John was a little hesitant, but Mission Control said, "Well, have at it." So we started jogging, and then I realized, "This is a big rock!"

We kept jogging and jogging, and the rock kept getting bigger and bigger and bigger. We were going slightly down hill, so we didn't sense how big it was at first, and so we get down to this thing and we called it "House Rock." It must've been 90 feet across and 45 feet tall. We walked around to the front side or the east side, which was in the sunlight, and, you know, it was towering over us. And we had this little hammer in our hand, thinking, "What are we going to do with this rock?" When we got down there—this humongous rock towering over us—John and I hit it with a hammer and a chunk came off. We were able to collect a piece of House Rock. Then we had a struggle hiking back, uphill.

We had some nuisance things that happened. Sample bags falling off and clips not working. But all and all, everything worked right.

Astronauts, even to this day, will talk about how difficult it is to work in a full pressure suit.

You didn't really sense a problem. I mean, you're so pumped up out there on the lunar surface. We had a good cooling system that kept us very comfortable. We had a bottle of water Velcro'd to the inside of the suit that we could drink out of. A little high-energy food bar was also Velcro'd inside the helmet that you could snack on to keep you nourished.

But you couldn't scratch your nose.

You couldn't scratch your nose. But you could reach over and hit the side of your helmet and things like that. Or if you had to, you could sort of rub your back against the suit. You couldn't bend over at the waist without great difficulty. You couldn't really bend at the knee so you had to learn how to operate the suit to make it work for you. It turned out that when we got back inside and took everything off, we were exhausted. I mean, it was hard work. You're squeezing that glove for 7-8 hours was like, if you can imagine squeezing a ball for 7-8 hours. And doing curls and stuff in the suit, and trying to make it work for you. It was real work.

I remember a couple of times the flight surgeon said, "Slow down. Your heartbeat's up to 140 a minute or so. We want you to rest." Generally, we rested in the car when we drove from point to point. But a couple of times they had to make me just slow down and rest. The suit was tough work.

I remember my arms were cramping and the end of the fingers, under the fingernails, were sort of black-and-blue from blood bruises that were a result of the Apollo pressure suit. It kept you alive, though. Only one time did I have a feeling that, "I'm in trouble" in the suit, and that was the final part of our stay on the Moon.

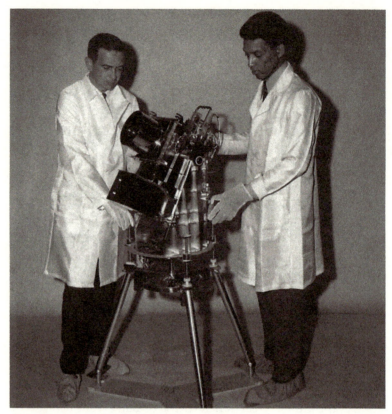

Above: Dr. George R. Carruthers, right, and William Conway, project manager at the Naval Research Institute, examine the lunar surface far-ultraviolet camera. (Photo courtesy Naval Research Laboratory.)

Next page:

Top right: Dr. George R. Carruthers, center, principle investigator for the lunar surface far-ultraviolet camera, discusses the instrument with Apollo 16 Commander John Young, right, and Apollo Program Director Rocco Petrone, left. Talking with Petrone is Apollo 16 Lunar Module Pilot Charlie Duke. (NASA Photo KSC-71P-544.)

Bottom right: Astronaut John Young, stands in the shadow of the lunar module behind the far-ultraviolet camera. This photograph was taken by Charlie Duke during the second Apollo 16 lunar surface EVA. Young set the prescribed angles of azimuth and elevation (here 14 degrees for photography of the large Magellanic Cloud) and pointed the camera. Over 180 photographs and spectra in far-ultraviolet light were obtained during the Apollo 16 mission showing clouds of hydrogen and other gases and several thousand stars. The United States flag and the lunar rover are in the left background. (NASA Photo AS16-114-18439.)

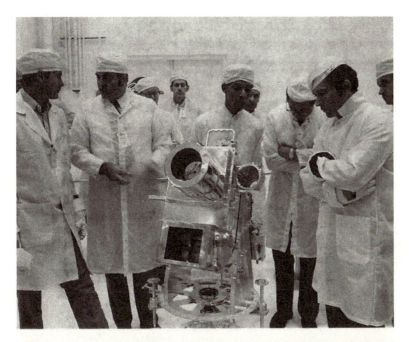

An artificially enhanced image of a 10-minute far-ultraviolet exposure of the Earth, taken during the Apollo 16 mission with a filter which blocks the glow caused by atomic hydrogen but which transmits the glow caused by atomic oxygen and molecular nitrogen. Note that airglow emission bands are visible on the night side of the Earth, one roughly centered between the two polar auroral zones and one at an angle to this extending northward toward the sunlit side of the Earth. (NASA Photo S72-40821.)

We were going to do the Moon Olympics, but John said, "Houston, we're going to do the Moon Olympics but we're running out of time, so we won't do that." And I said, "Yeah, Houston, I was going to bounce and set the high jump record." And I started just kidding around and bouncing, and when I jumped the last time, I went over backwards and disappeared behind the lunar rover, and the TV camera's pointed at me. That was a moment of panic . . .I was in trouble. You could watch me scrambling like that, trying to get my balance. I ended up landing on my right side, and bouncing on to my back. And my heart was just pounding, you know.

What was your concern? That you'd damaged the suit?

That I'd have damaged the suit. You know, the backpack is very fragile. I thought the suit would hold, but the backpack, with the plumbing and connections and all, if that broke, it was just like having a puncture in the suit. And, you know, falling over backwards on the Moon. Hardly ever did we think about, "We're in a vacuum. This thing's got to work." I don't remember thinking or worrying about that. But this time, as I started over backwards, the thought occurred to me, "I'm in trouble." I was able to spin right before I hit, and my right foot and right hand hit, and I bounced on to my back. And John came over and helped me up. I got real quiet, and you could hear the pumps running in the backpack. And I checked my suit pressure. It was okay so this fear began to subside.

Then I realized the TV camera was pointed at me and then embarrassment came. You know, "The stupid stunt." I forgot what Mission Control said. I think Tony England was our Capcom, and he said, "That's enough of that, guys." Anyway, that ended our Moon Olympics. Other than that time, there was not another moment where we didn't feel secure in our suits.

Did you find that you adapted very quickly to moving in one-sixth G.

Yes, we did. I found that either the hop or the skip was the best for me. John was more of a jogger-type. I put my right foot out front and I just sort of skipped along. Then again, if the ground was level, you would start a little jog and it was sort of effortless as you went across. We fell down a lot, at least I did. I found when you did fall down on your front, then you could just do a series of push-ups and you'd sort of rock yourself back up and then eventually you'd pop up.

It was great fun. John and I really had a ball. We were joking and just having a tremendous sense of enjoyment and adventure.

It was a real adventure for me and John and it built a friendship that is real solid now, 27 years later.

If you had a chance to go back to Descartes, what do you think you'd find? Do you think your lunar module and the flag and all would still be there?

Without a nearby meteorite impact, I believe it'd still be there.

You don't think the flag blew down when you took off?

Well, it didn't look like it. I was running a camera out of the window and as we pitched over it [the flag] wiggled a lot, but I think it stayed upright.

One of the things that changed on your mission from the previous one was a reaction, I wonder if you think perhaps it was an over-reaction, to the exhaustion and the irregular heart rate that Jim Irwin experienced on Apollo 15 because of the heavy workload. As a result, they really loaded you guys down with potassium.

I'm glad they did that. They changed our medical kit. They gave us some sort of injection that we could take if we did see heart problems developing. That was an experience that was (looking back), at the time, humorous. We were in flight medicine getting a briefing on this new medical kit with this thing that was heart medication. It wasn't oral, but it was an injection. One of the flight surgeons was telling us, "Well, you count down so many ribs" and this was going to be injected right into the heart muscle. And you'd press it. (It was like those battlefield syringes that would fire this needle in and inject the stuff.)

So he said, "Well, let me demonstrate this." And he took a Styrofoam ball about the size of a grapefruit and he pressed this thing. When it went off, the Styrofoam ball exploded and I almost passed out. I knew at that moment, "that thing is never going into my heart and I don't care how sick I am." We never had to use it, of course but they did put the potassium in to try to regulate the heartbeats.

We also generated tools that would help us overcome the hard workload previously experienced on Apollo 15. For instance, the core that we had to drill (10 feet deep), [Apollo] 15 could hardly get the thing back out of the surface. Dave Scott and Jim Irwin (Dave Scott was strong as a gorilla) had a tough time so we developed this little jack. It was a collar that went around the stem of the drill that was sticking out of the surface. Like a car jack, I could just keep jacking it and this would slowly jack this thing out of the surface. It was easy to do. Those kinds of things came about in debriefing.

We found that potassium did work as a laxative and we had problems with, you know, our BMs[5]. At least I did and it turned out that it wasn't very pleasant. The Apollo system wasn't the most high-tech system in the world to use. While we didn't have any real serious accidents, it was just the frequency . . .

You mentioned that at the end of the day, you get back into the lunar module, get the suits off, and you're in an environment one-sixth gravity. You have no place to lie down. How were you able to sleep between EVAs?

You're tight, but I wouldn't call it "jammed in." Once we got off the suits, there was a place between us, behind us over the ascent engine cover, where we could drape the suits over and they were out of the way. For me, I could lean back and sort of semi-sit on the environmental control unit. We just felt it was comfortable, really, without having to sit down.

When we got ready for our rest period, we each had a little beta-cloth hammock. Mine attached across the LM. There were hooks on the left side, and two hooks on the right side. So I hooked up this little hammock, cinched it up, so it was about six inches off the floor, and I rolled up one of the liquid-cooling garments for a pillow. John's was up above the ascent engine cover. He hooked on to the sides of the instrument panel and the back bulkhead, so we were sort of in a cross. He would climb up and get in his hammock. Amazingly, he went right to sleep.

The first night, we had changed our flight plan due to this late landing that required us to go to sleep before we went outside for the first time. I'm on the Moon and six hours after landing, "go to sleep." My mind's just racing like crazy and, even though we were tired, I couldn't get to sleep. So I asked Mission Control if I could take a sleeping pill, which I did, and then I drifted off to sleep. And the sleep was very comfortable in the little hammocks.

I remember that first sleep period. We had been warned that, some time during the night, a master caution alarm would go off due to a reaction control regulator problem that we had prior to descent. When we powered up the system, we lost the pressurization controller; both regulators failed and we were overpressurizing, but John quickly vented the pressure into the ascent engine tank to save the mission. As we rested, it began to heat up and overpressurize again. And we knew that when we got to a certain level, this warning would go off. Sure enough, I've just dozed off and "bong, bong, bong!" I have the headset on and this master alarm went off. I almost went through the top of the lunar module. I cycled whatever I had to cycle to vent the pressure and went back to sleep.

It was all right for the rest of the time?

It was all right the rest of the time. After that first night, I didn't need any sleeping pill. I was physically tired after working on the lunar surface and able to get right to sleep. I'd say we averaged, maybe, six hours sleep on the lunar surface. The best rest I got was on the way home. You started home and your attitude was, "Mission complete. We've done it. We're on our way back. Let Mattingly handle this thing from now on."

On the way home, I was being monitored, and the flight surgeon was watching me sleep. I think my heartbeat got down to, like, 28. I was dead to the world, and it was really refreshing. When I woke up, they said, "Man, we thought you were dying."

While you were on the Moon, Tony England, your Capcom at Mission Control, relayed the news to you that the House of Representatives passed the NASA budget with funding to go design and build the Space Shuttle.

Charlie Duke took this photo of a family picture that he left on the surface of the moon. The photo, taken by Loudy Benjamin, is shrink-wrapped and contains a message on the back which reads "This is the family of Astronaut Duke from Planet Earth. Landed on the Moon, April 1972." Underneath the message are the signatures of his wife and kids. (NASA Photo AS16-117-18841.)

That was exciting. In fact we had just saluted the flag. John had just saluted the flag and gave that little jump for joy and saluted. When we were changing positions, if I remember correctly, right after that Tony relayed that they had just received word of funding for the Space Shuttle. And John and I were excited. John made the comment, "Yeah, we really need that Shuttle."

Two other things that I thought were memorable while I was on the Moon. I was the only Air Force officer on the Moon during 1972. We had two missions, Apollo 16 and 17, and it was the 25th anniversary of the Air Force in 1972. They had some special medals struck, like silver dollars, that had the Air Force seal on one side and Apollo on the other. I took two to the Moon with me, with the approval of NASA, and left one on the Moon (took a picture of it) and brought the other back. So I was able to say "Happy birthday, Air Force," while we were on the lunar surface. I also had an Air Force flag that I took and gave them a piece of Moon rock. These are in a museum now up at Wright-Patterson Air Force Base.

The other was: I took a picture of my family. Our kids were five and seven. I had a little picture that had been taken in my backyard by one of the NASA guys, Loudy Benjamin, and we had that shrink-wrapped. On the back of this photograph we wrote: "This is the family of Astronaut Duke from Planet Earth. Landed on the Moon, April 1972." And the kids signed it to get them involved with the flight. I left that on the Moon and took a picture of the picture, and that's one of our neatest possessions now.

ENDNOTES

1. One of the most serious threats to the landing came in the form of a "1201" alarm from the lunar module's (LM) computer. When it flashed those numbers, the LM was signaling that it had an overload of information and needed to sort through its tasks. A continuous alarm would have meant an aborted landing, but its intermittent nature in this instance led the LM computer expert Steve Bales to signal a "go" for landing.

2. "Mascons" were unexpected irregularities in the gravitational field of the Moon, discovered by the unmanned lunar orbiter mission in 1966-67.

3. In 1963, Bell Aerosystems Company in Buffalo, New York, was initially awarded a contract by NASA to design and develop two unique manned Lunar Landing Research Vehicles (LLRV). Bell based their design for the LLRV on the so-called "flying bedsteads" that were flown in the 1950s to study the potential of jet lift for vertical take-off and landing (VTOL) aircraft. On May 6, 1968, Neil Armstrong, while making his 21st flight in the LLRV, narrowly escaped disaster. After lifting off and flying to an altitude of 150 meters, he began a simulated lunar landing. Suddenly the LLRV began to pitch forward at about 70 meters while at the same time picking up speed. Armstrong attempted to make corrections but could not stop this motion and chose to eject. Seconds later, the LLRV crashed in a field while Armstrong parachuted to safety. After further initial tests of the first two vehicles at NASA's Flight Research Center (now the Hugh L. Dryden Flight Research Facility), three improved versions of the trainer, renamed the Lunar Landing

Training Vehicle (LLTV) were ordered by the Manned Spacecraft Center. Two of the LLTV's were used by astronauts at the Kennedy Space Center while one remained at Ellington Field in Houston. After a second near fatal mishap occurred when test pilot Joe Algranti punched out of LLTV No. 1 at Ellington Field, there was talk of ending the flights altogether. The program remained, however, after astronauts argued that the vehicles accurately forecast the LM's handling characteristics during the final critical moments of descent to the lunar surface. Ed Hengeveld, "Training for a Lunar Landing: The LLRV and LLTC," *Quest*, Vol. 6 No. 1, pp. 50-54.

4. NASA's Langley Research Center in Hampton, Virginia, developed a Lunar Landing Research Facility (LLRF) which was basically a tethered version of the LLRV, suspended from a giant support structure, that allowed an LM-type vehicle to practice the final 50 meters of lunar landing while hanging from moving cables. The LLRF was primarily used by Langley pilots to test instruments and software designed for the LM. The original helicopter-like cockpit was later modified to allow the pilot to stand upright, as he would during a real lunar landing, so he could get used to flying in that unusual position. The LLRVs and LLTVs did not allow a pilot to practice while standing up because they were equipped with ejection seats. Ed Hengeveld, "Training for a Lunar Landing: The LLRV and LLTC," *Quest*, Vol. 6 No. 1, p. 51.

5. The potassium that the crew received during their mission was in the form of enriched orange juice. The crew had some pretty strong words to say regarding this mixture. Prior to the first lunar EVA, and during their first meal on the Moon, Duke managed to accidentally release some orange juice from the siphon attached to their space suits. The juice eventually worked its way into his hair and helmet, causing him some distress. At one point he remarked, "I wouldn't give you two cents for that orange juice as a hair tonic." While the incident probably amused Duke's fellow moonwalker at the time, the orange juice would later come back to haunt John Young. As Duke and Young were debriefed after returning to the LM from the first EVA, Young left his microphone on voice activation, meaning that every time he spoke, it was relayed live to Houston. This resulted in an unusually vivid and colorful account of post debrief activities. Houston picked up quips from Young about how the potassium enriched orange juice was giving him gas, including a few colorful remarks about never eating oranges again. The Capcom on duty picked up on the conversation and politely informed the crew of the "hot" mike. Richard S. Lewis, *The Voyages of Apollo: The Exploration of the Moon* (New York: The New York Times Book Co., 1974), pp. 245-248; For a transcript of the infamous conversation, see Mission Elapsed Time 128:50:37 - 128:54:15, in Eric M. Jones, ed., "Apollo 16 EVA 1 - Debrief and Goodnight," [Apollo Lunar Surface Journal Home Page]. [Online]. Available: http://www.hq.nasa.gov/alsj/a16/a16.eva1debrief.html.

Apollo 17 mission commander Eugene A. Cernan, and lunar module pilot Dr. Harrison H. Schmitt at right, familiarize themselves with equipment used in the lunar module while undergoing prelaunch training in the lunar module simulator at the Flight Crew Training Building at the Kennedy Space Center on October 27, 1972. (NASA Photo 108-KSC-72PC-541.)

CHAPTER 12

HARRISON H. SCHMITT

(1 9 3 5 –   )

Born July 3, 1935, in Santa Rita, New Mexico, Harrison Hagan "Jack" Schmitt became the first and only geologist to land on the Moon when he served as lunar module pilot of Apollo 17, the last Apollo lunar landing.

Growing up in Silver City, New Mexico, Schmitt was exposed to the sciences, particularly geology, at an early age while assisting his father, a mining geologist, during weekends in the field. While attending the California Institute of Technology, Schmitt was intent on becoming a physicist but soon found that he made a much better geologist and switched majors. After graduating from Cal Tech in 1957, Schmitt went on to obtain a PhD in Geology from Harvard University in 1964.

After finishing a postdoctoral fellowship in 1964, Schmitt joined the recently established Astrogeology branch founded by Eugene Shoemaker at the United States Geological Survey (USGS), in Flagstaff, Arizona. Schmitt helped develop techniques for field geology on the Moon and was named project chief for lunar fields geological methods. Soon after, NASA announced that they were seeking applicants for scientists-astronauts and Schmitt promptly applied.

NASA's announcement for applicants for this new group of astronauts came after considerable pressure from within the scientific community who pushed to have a trained scientist crew member onboard one of the Apollo lunar landing missions. The National Academy of Sciences acted as a screening board, sending their recommendations from those applicants received to NASA. From the 1,000 applicants, 16 names were sent to NASA, including Schmitt's and two others from the USGS astrogeology branch. Schmitt was rejected along with the two other USGS applicants. Shoemaker was shocked by this and personally intervened on Schmitt's behalf to help get him selected. When NASA announced the six new scientist-astronauts to the public in June 1965, Schmitt's name was among them.

Schmitt immersed himself in his work, having no life outside of the Apollo program. One of Schmitt's first assignments after earning his pilot's wings and completing basic astronaut training was in working on the Apollo Lunar Science Experiments Package (ALSEP). He soon became the bridge between the astronaut corps and the scientific community, earning the respect of both. He worked successfully to educate his astronaut colleagues in the area of lunar geology, beginning with the crew of Apollo 8. In the summer of 1969, Schmitt helped recruit Lee Silver, a well-know field geologist and geochemical analyst, to help in training the astronauts in lunar surface field geology.

Schmitt also pressed for a chance to fly on his own mission for he knew that as a geologist, he had an edge over the other scientist-astronauts. In addition to knowing the science, Schmitt also had to master the many other technical aspects of a mission so he applied himself to the simulators, getting in as many hours on the different equipment as possible. His hard work eventually paid off as he was assigned to be the lunar module pilot for Apollo 17, displacing Joe Engle on the last lunar landing mission.

Schmitt, Commander Eugene Cernan, and Command Module Pilot Ronald Evans were launched in the early morning of December 7, 1972, the fiery climb of their Saturn V lighting up the Florida coastline for hundreds of miles around. After reaching lunar orbit three days later, Schmitt and Cernan left Evans in the command module "America" and boarded the lunar module "Challenger" to descend to the Sea of Serenity near the crater Littrow and the surrounding Taurus Mountains.

In July 1973, following Apollo 17, Schmitt was named a Sherman Fairchild Distinguished Scholar at Cal Tech, an appointment that ran until July 1975. From February 1974, Schmitt served as chief of the *scientist-astronauts within the Astronaut Office in the Flight Crew Operations Directorate at the Johnson Space Center. In May 1974, he became the assistant administrator of the Office of Energy Programs at NASA Headquarters.*

After retiring from NASA in 1976, Schmitt went on to pursue a career in politics. That same year, he ran for and was elected to the U.S. Senate for the State of New Mexico. After an unsuccessful 1982 bid for reelection, Schmitt became an independent consultant for science, technology, and public policy.

Editor's Note: The following is an edited transcription from a presentation given by Harrison Schmitt as part of an Apollo 11 20th anniversary celebration speakers series held during the week of July 17-21, 1989, at the Johnson Space Center in Houston, Texas. This presentation entitled "The Moon Before Apollo" was given on July 17, 1989, followed by a question and answer period from the audience.

The Earth is the basis of the knowledge with which we began our specialized and detailed studies of the Moon. Geology and Earth Sciences, in general, form the basis of those analyses through, not only decades before Apollo, but literally, centuries. We had a little bit of information about the Moon from this perspective as we began. Science may have been the least of the reasons most of the nearly 500,000 Americans signed up for Apollo. However, along with technology, national confidence, and human spirit, science benefited permanently from our exploits. Remembering just how little we knew about the Moon before Apollo serves to emphasize both how far we have come and how far we now may go. The Apollo program gave us a first-order understanding of the sequence of events and processes by which the Moon became what existed three billion years ago, and what remained until Armstrong, Collins, Aldrin, and myself arrived. However, this conclusion relates not only to events and processes after the Moon's formation near the Earth 4.6 billion years ago. I am one of the few remaining skeptics concerning the "Mars-size asteroid collision with the Earth" theory. The Moon's planetological tail, like that of my golden retriever, Pecos, wags on much fact but also on much uncertainty. The fact we have seen and measured; the resolution of uncertainty, we must leave to the future.

The first known facts about the Moon were accessible even to our distant ancestors as they planned monthly hunting and gathering around ancient African lakes. Had they looked closely at the full Moon, as they may well have done, and I'm sure did, they would've seen very light regions interrupted in some places by sharply defined circular, or nearly circular, areas. These circular areas and other more irregular regions would have been seen to be very dark, relative to the light regions. If our great ancestors had good eyes and squinted, they would have seen even then, several very bright spots on both the light and dark regions.

The Apollo 17 Saturn V (Launch Vehicle SA-512) during its August 28, 1972, move on Mobile Launch Platform 3 from the Vehicle Assembly Building at the Kennedy Space Center to Launch Complex 39A. (NASA Photo KSC-72PC-430.)

In the 16th century, the observations and logic of Copernicus, following in the footsteps of Aristarchus and probably others before him who remain unrecorded, demonstrated that the Moon orbited the Earth while both bodies orbited the Sun. Then, in the 17th century, Galileo invented the telescope. Through this remarkable example of human ingenuity and technology, it was soon discovered that the very light areas were cratered terra, or highlands; the dark areas were less heavily cratered basins, or lowlands—at first mistaken for seas and called maria—and the bright spots were craters with vast systems of bright radial patterns, or rays. Kepler, Newton, and others, also in the 17th century, provided the mathematical means of calculating the Moon's shape, roughly a sphere; size, a radius of about 1,738 km; a mean density of about 3.34 g/cm3; and a moment of inertia, about .400 for the principal moment. These facts, although slightly inaccurate, place limits on the composition—that is, the overall Moon was similar to stony meteorites that fall on the Earth and had significantly less iron relative to the Earth and the Sun. They also placed limits on the distribution of mass within the Moon. With any logical bent, these same observers must have realized that the Moon had no clouds or other apparent protection from the ravages of space and the Sun, although I'm not aware of that observation being recorded.

Gilbert's studies of lunar surface features in the 19th century began modern geological considerations of the Moon. During the mid-20th century, Yuri, Keifer, and Baldwin contributed new insights from indirect analysis on composition, age, shape, motions, and surface properties. In the years just preceding human footsteps on the Moon's surface, and as it became apparent that humans might soon go there, knowledge of this small planet's past expanded rapidly through specialized investigations. Schumacher, Hackman, and many others used telescopic observations, photographs from space probes, and the extension of Steno's law of superposition to add refinements to the more ancient text about the Moon. That law of superposition, by the way, merely states that things that are on top are usually younger. There are, I'm sure, exceptions to that. The highlands appeared saturated with large meteor-impact craters, at least 50 km in diameter, and were clearly the oldest visible crust of the Moon. The analyses by the last Surveyor spacecraft demonstrated that this light-colored crust contained significant calcium and aluminum. The large, circular basins not only cut through the upper crust, but through large, irregular basins as well. Evidence existed for previously unrecognized materials that formed light-colored plains in old basins and lowlands. Early in this modern, but still remote, study, it became clear that the dark maria were vast plains of remarkably fluid volcanic flows, visually similar to the rock basalt common here on Earth. Both Ranger photographs and Surveyor analyses confirmed these conclusions. These flows filled the large circular and irregular basins on the Moon's front side and covered some of the light plains as well.

The Lunar Orbiter photography showed that the back side of the Moon had only local areas of dark maria to interrupt an almost continuous expanse of cratered highlands. Tracking of Lunar Orbiter and, later, the Apollo 8 and 10 missions, disclosed the unexpected presence of mass concentrations [mascons] in the large circular basins, indicating great structural strength—at least at present. It soon became clear that little major activity had occurred on the Moon's surface after the last mare flow had cooled, estimated from crater frequency assumptions as having occurred a few hundred million years ago. This estimate was only off by a factor of 5 or 6.

As new studies became more detailed, an overprint of subtle features became apparent. The surfaces of the maria and highlands had been altered at some later, but as yet undetermined, time. A lengthy ridge and volcanic island system splits the Moon's largest western mare basin, Procoleram; strange, sinuous rills and light-colored swirls locally interrupt the surface of many areas; and extensive, relatively young, extensional fault systems—that is, pull-apart fault systems, in particular what we called "robins" —cut large regions of the surface. These fault systems showed that the outer crust of the Moon could not support significant shear stress, even though it could support the extensive mass concentrations. Overall, a pulverized rock layer, or soil, or regolith covered essentially all the surface, attesting to continuous bombardment from space and radiation from the Sun. Before any pieces of rocks or scoops of soil arrived from the Moon, we knew that the following sequence of events had occurred once the Moon had reached roughly its present size: An old, highly impacted crust—rich silicate crust—rich in calcium and aluminum, had formed. Second,

Harrison Schmitt shares a moment of relaxation with astronaut Alan Shepard during prelaunch suiting operations at the Kennedy Space Center. (NASA Photo 108-KSC-72P-547.)

large irregular, and, later, circular basins were created in this old crust by very large, almost inconceivably large, impact events. The floors of many basins were covered, in part, by light-colored, smooth plains of unknown origins. Most of the basins were filled to unknown depths by basaltic lava flows. After all major changes such as those I've mentioned had ceased, the surface was slightly modified by the formation of constructional volcanic deposits—

faults, rills, and light-colored swirls. And finally, pulverized rock soil, or regolith, formed in response to meteor and radiation bombardment from space and the Sun.

What we did not know were the how and when of the Moon's formation and evolution. Scientists still struggle with the "how" of the formation of the Moon. We may still be barking up the wrong tree on that one, or, as we say on the Moon, "wheezing up the wrong crater." However, I cannot resist this opportunity to summarize the evolutionary sequence I find supports the data of which I am aware at the conclusion of the Apollo explorations. The major lunar evolutionary stages and their defining characteristics appear to be as follows.

First, the beginning—clearly about 4.6 billion years ago in which we had the formation of the Moon roughly contemporaneously with the formation of the Earth. The process of formation cannot have produced a fully molten or partially molten Moon because subsequent differentiation would have eliminated both, one, the relative undifferentiated material rich in volatiles and primordial lead required as a deep-source material for the orange and green soils, about which many of you have heard; and, second, the density reversal at 200 km or 300 km depth required to maintain the present average density of the Moon—two, as I see it, very important constraints on any model of formation. This suggests a beginning through accretion of relatively cold material rather than through catastrophic separation of a preexisting and largely differentiated Earth.

Secondly, we had the melted shell. Now we have somewhat more agreement that, between 4.6 and 4.5 billion years ago, accretionary melting, or residual heat from formation, and volatile depletion of the outer 200 km to 300 km of the Moon, so-called magma ocean, with accompanying differentiation into a 60 km to 70 km thick anorthocitic crust and a 200 km to 300 km thick mafic, or iron- and magnesium-rich upper mantle. It is likely that, prior to the development of a coherent crust, large accretionary impacts mixed undifferentiated melt into the protocrust, or early crust, adding both an overall gabbroic contaminant (that is, a magnesium- and iron-rich contaminant) and many splash intrusions of significant size represented by the so-called INT suite of lunar samples. The residual liquid in this differentiating melted shell gradually took on the chemical characteristics attributed to an original type of rock that we've never really seen called "URKREEP." KREEP, as you may recall, was the acronym applied to potassium, rare Earth element, and phosphorus rich samples that reached prominence mainly in Apollo 14. The original material from which KREEP was evolved has taken on the name "URKREEP." Due to the concentration of radioisotopes, it probably remained a liquid, radioisotopic heat maintaining it as such. During the accretion and melting of this shell, emissible iron sulphur liquid settled to its base and ultimately migrated downward to form a still fluidized core of about 500 km radius. These processes of differentiation, however, did not significantly affect the composition of relatively undifferentiated lower mantle material.

The third phase, the cratered highlands: 4.5 to about 4.3 billion years ago, saturation of the lunar crust, formed and preserved impact structures, 50 km to 100 km in diameter, and created a remarkable time in lunar history as well as, almost certainly, in Earth history. Significant regional homogenization of the upper crust and intense brecciation of the lower crust to at least 25 km depth took place at this time. The insulating properties of the brecciated and pulverized (that is, intensely broken) upper crust allowed the gradual accumulation of radioisotopic heat necessary to eventually partially remelt the mantle and produce the basaltic maria.

The fourth phase, the old large basins and crustal-strengthening phase: about 4.3 to around 4.0 billion years ago, there was the formation of what are called pre-nectarian large basins (that's just a generalized code for the age) with rapid isotopic adjustment of the crust. This was permitted by both the URKREEP—the original KREEP liquid—underlying the crust and the pervasive fracturing in that crust. After the first, now somewhat irregular, large basins formed over much of the Moon's surface, the relatively low density URKREEP liquid moved upward into the deeply and closely fractured crust. KREEP model ages of 4.4 to 4.3 billion years may indicate the time of the initiation of this movement. Moving upward into the cooler upper crust and becoming significantly contaminated with crustal debris in the process, the KREEP-related liquid ultimately crystallized and formed interlocking networks of intrusions. KREEP-basalt crystallization ages of 4.2 to 3.9 billion years indicate the timing of the solidification of these intrusions. Once solidified, these interlocking intrusions—and I think you can imagine what that would be like in a broken crust like the Moon's—these interlocking intrusions combined with the removal of the underlying destabilizing URKREEP liquid, strengthened the crust, which had to happen. KREEP-related magmas contaminated with anorthocitic crustal materials may have reached the floors of the deepest early basins. Later, redistribution of materials from near-surface KREEP basalts, by impacts, may help to account for the observed age spectrum of the KREEP basalt samples we have found throughout the regolith.

The fifth phase, the young large basins: in a brief period, probably about 4.0 to 3.9 billion years ago, there was the formation of what are known as nectarian and embriem age large, circular impact basins in a crust strong enough now to support mass concentrations and, indeed, mass deficiencies indefinitely, until we began to observe them in this century. Crustal penetration of the embriem event was deep enough to eject significant volumes of KREEP-related intrusive material around its basin (where we now see it).

The sixth phase, the light colored plains: this is a very controversial area and has been for many decades among those who have studied the Moon; but it does appear, both from Earth-based and satellite-based mapping and from visual observation of the surface by myself and a few others, that about 3.9 or 3.8 billion years ago, lunar-wide debris was deposited in most of the then existing basins and may have resulted from a combination of the final degassing of the mantle that preceded the formation of the basalt maria (magmas), something that one would expect to happen, with crustal debris entrained as the gasses moved upward and the wide-scale redeposition of fine debris ejected during large basin formation. Many of the observed surface expressions of the so-called light-colored plains, or calix formation, and the nature of the samples from the Descartes landing site, appear to be explained by this mode of origin.

295

The seventh period, the basaltic maria (the man-in-the-Moon phase): from 3.8 to 3.0, and maybe more, billion years ago, there was surface eruption and subsurface intrusion of basaltic maria, probably appearing first at a salinographic low in the vicinity of northern Tranquilitatus and southern Serenitatus with subsequent eruptions appearing in very roughly concentric zones around this low. Rare Earth element geochemistry suggests that progressively deeper zones of the lunar mantle melted to give rise to the age sequence and, of course, chemical sequence that we see in the mare basalt magmas. Pyroclastic eruptions (that is, fire-fountain type eruptions) of basaltic material—such as the orange and green soils—containing volatile components and primordial lead derived from below the original melted shell accompanied late-stage mare eruptions in many, if not most, regions.

The eighth and final phase of lunar evolution, the changing crust: from 3 billion years ago to the present, there have been a number of changes to this crust, none of them extremely obvious but still changes that we now know about. The formation of the Procolarem volcanic ridge system and the eastern limb and western far side light-colored swirls are most important, and they are possibly due to a brief period—a very brief period —of mantle convection which, because of the Moon's size, was turned off quite quickly. The present lunar regolith formed during this period in response to a continuous, but gradually declining, meteor impact frequency.

Apollo 11 began the process of understanding the evolution of the Moon and, indeed, the early evolution of the Earth. From the Moon, we gained information about the early history of the Earth that would have been impossible to gain from the Earth itself. Through this remarkable effort on the part of national leaders, managers, engineers, workers, industry, a worldwide community of scientists, and the American people and their families, and through the other Apollo and lunar missions, we now have looked with new insight at our own planet and other terrestrial planets. Because of this, investigations reached towards real understanding of the early differentiation of the planets, the nature of their internal structure, the environmental dynamics during the origin of life—one of the more exciting aspects of it—the influence of very large impact

Astronaut Gene Cernan, Apollo 17 commander, salutes the deployed United States flag on the lunar surface. The lunar module is at left background and the lunar rover also in background, is partially obscured. The photo was shot by Harrison Schmitt. (NASA Photo AS17-134-20380.)

events, and the effects of early partial melting of protomantles, or early mantles, of the planets, and possibly the earliest manifestations of plate techtonics. The Apollo explorations were of incalculable value in adding the reality of known materials and processes to the interpretation of data from subsequent automated explorations of the solar systems—something we also, I think, tend to forget.

What would we have known about Mercury and Mars and Venus and other terrestrial-type planets without the detailed understanding we achieved about the Moon? Now, it has been recently noted that, early in the third millennium, the Moon's mare regolith may provide vast and environmentally benign energy resources required by Earth and the consumables required for martian settlement. This prospect of the next great human challenge and the next great human adventure being tied together by the resources of

A close-up view of the lunar rover at the Apollo 17 landing site Taurus-Littrow. Note the makeshift repair on the right rear fender. During EVA-1, Gene Cernan's rock hammer, its handle sticking out of his spacesuit pocket, snagged and broke off part of the fender. Minus a fender, this soon created a problem as lunar dust went flying everywhere while the rover was in motion. Following a suggestion from Astronaut John Young in the Mission Control Center at Houston, the crewmen repaired the fender early in EVA-2 using lunar maps and clamps from the optical alignment telescope lamp. Schmitt is seated in the rover while Cernan took this photo. (NASA Photo AS17-137-20979.)

the Moon, I think is extraordinarily exciting. Not too bad for a program born of international competition, but made more than that by the early and expansive insights of those who designed and operated the systems necessary to win that competition. Science and humanity owe these giants more than will ever be repaid.

What potential does the Moon have as an energy source for use here on Earth and as a source of materials to assist us in the further exploration and settlement of space?

I think there is a lot of interest in that and more, probably, is deserved—principally because the greatest danger that this thin shell of an atmosphere or biosphere (the Earth) has in the foreseeable future probably comes from the continued consumption of fossil fuels. And so, some alternative in the not too distant future to fossil fuels for both the generation of electricity and portable fuel that we use in transportation is essential. We must look not only here on Earth but elsewhere for such alternatives. Looking at the Moon for such an alternative, we see basically three possibilities, two of which are closely related. One is to utilize solar energy, either collected on the Moon itself, utilizing lunar materials for the collectors, or in solar satellites, utilizing lunar materials for the collectors and beaming that energy back to Earth in a network of power beaming and distribution to be utilized essentially anywhere that it's needed on Earth. There are, as you might imagine, tremendous complexities with that idea. But the cost of doing that and the relative return on any investment that might be made is not clear and won't be for some time, but it certainly deserves some study.

The other alternative, which I personally find even more exciting, is that we have found in the Apollo 11 samples the presence of trace amounts of an isotope of helium called Helium 3—10 to 20 parts per billion of this particular isotope of helium. Helium 3 is a potential fuel when combined with dueterium (an isotope of hydrogen) in fusion reactors. You probably have heard of other fuels for fusion reactions—dueterium can be a fuel as well as tridium, which is the current most extensively researched fuel for fusion reactors—but tridium and dueterium have the difficulty of producing very long-lived radioactivity in their reactors. Tridium has some short-lived radioactivity, of course; but the problem is that, when you burn them in reactors, you get a lot of high-energy neutrons and they, in turn, create radioactive contamination within the reactors themselves, and, indeed, actually damage the reactors. Whereas, with Helium 3, the principal product of fusion is a proton, and a proton is relatively benign with respect to creating radioactivity. There are some side reactions that produce some neutrons, but they're not in major amounts, as it now appears from the calculations. Not only are protons attractive from the point of view of ultimate radioactive waste, but they're attractive in that they represent a way of directly converting energy to electricity. So, the efficiency of Helium 3 fusion reactors may well be as much as twice that of other types of reactors. Now, you may ask, "But, 10 or 20 parts per billion in the lunar soil? How in the world does that make any sense as a terrestrial energy source?" Well, let me give you a couple of numbers to put that into perspective. A metric ton of Helium 3 contains as much energy as two billion dollars worth of coal burned today as a source of electricity. Two billion dollars worth of coal will provide the electricity (in this country) for 10 million people for about a year. As an old economic geologist, when I hear about a resource that may be worth two billion dollars a metric ton, I can start to think of all sorts of ways that I might get access to that resource. It doesn't mean we've figured it out yet, but I think it does serve to illustrate why the establishment of a lunar base in the perspective of protecting our global environment starts to make an awful lot of sense as a national investment in the future. And, you get a tremendous payoff in addition, because the next great human adventure will be the settlement of Mars, and in the process of refining and processing lunar soils for Helium 3, you create vast tonnages of other gaseous elements: hydrogen, water, carbon compounds, and nitrogen compounds in particular. Those are exactly what one needs to initiate and catalyze the settlement not only of the Moon but of Mars. So, the Moon acts as a bridge between the great challenge we face here on Earth and the great adventure we have ahead of us in the solar system.

Looking back, how valuable were the unmanned precursors to the Moon such as Ranger and Surveyor, to the manned Apollo missions?

In advance of Apollo, given the state of our knowledge and the state of consensus about the surface of the Moon in several respects, Ranger and Surveyor were of tremendous importance. Just as an anecdote: My mother, who is 85, still remembers more vividly seeing Ranger pictures on television with "Live from the Moon" printed out at their base than, I think, she does the Apollo 17 mission to the Moon. Someone born in 1904 seeing something on television in the first place has an interesting perspective, but seeing a picture that, believably, is live from the Moon has an even greater perspective. And, I wouldn't be surprised if she's not alone, that the first clear pictures that were coming from the Moon, just prior to the impact of the Ranger vehicles, did a lot to lay a psychological base within this country for us to go on and do other things. In addition, having done my first exercise in lunar traverse planning on a Ranger 8 photograph, I think it taught us a great deal about the complexities of doing geology on the Moon in that, the closer you got to the Moon, the more difficult it was to decide what are we were going to do to maximize the efficiency of our exploration activities? At least it taught me that lesson, and, with being deeply involved in the traverse planning later on, I think it was an important experience for me.

There was a more political question than I think a geological question about the nature of the surface. There was, I think, a geological consensus that astronauts would not be swallowed up by dust on the surface. There was a misinterpretation of some radar data by Dr. Tommy Gould that he parlayed into a great deal of excitement and expense for NASA in suggesting that there was a thick layer of dust on the Moon. The Surveyor spacecraft, of course, by landing, totally disproved that, and that should've been no longer a concern. It didn't quiet Tommy very much, but it should've been of no great concern. In addition, Surveyor began to outline the broad chemical parameters that we were going to deal with in our analysis of the lunar samples. And even though a great deal, if not most, of the groundwork for the analysis of the lunar samples had already been laid—internationally, by the way—and it's important to remember that the analysis of lunar materials was really the first major international science program in space. We often forget that science had internationalized space long before Apollo-Soyuz did. Nevertheless, Surveyor began to alert these people to the kinds of systems they were going to be dealing with: in the highlands—silicate systems dominated by calcium and aluminum—and in the mare—silicate systems dominated by magnesium and iron. And that, again, I think was an important

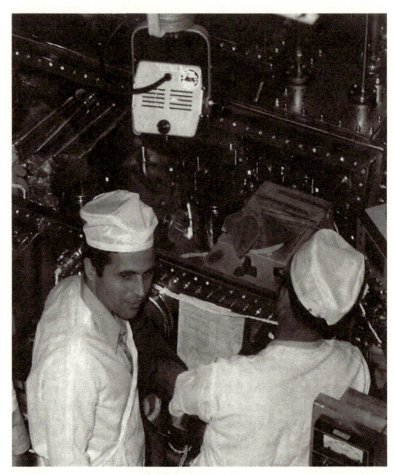

Scientist-Astronaut Harrison Schmitt (facing camera), was one of the first to look at the sample of "orange" soil which was brought back from the Taurus-Littrow landing site by the crew of Apollo 17. Schmitt discovered the material at Shorty Crater during the second EVA. The "orange" sample, opened on December 27, 1972, is in the bag on a weighing platform in the sealed nitrogen cabinet in the upstairs processing line in the Lunar Receiving Laboratory at Houston. The sample was removed from one of the bolt-top cans visible to the left in the cabinet. Schmitt's first reaction was, "It doesn't look the same." Most of the geologists and staff viewing the sample agreed that it was more tan and brown than orange. Closer comparison with color charts showed that the sample had a definite orange cast. After closer investigation and sieving, it was discovered that the orange color was caused by very fine spheres and fragments of orange glass in the midst of darker colored, larger grain material. (NASA Photo S72-56362.)

criteria. And Surveyor, of course, also added to the psychological confidence that we could do it. Even though the first few didn't make it, the latter ones did, and we showed that, operationally, working precisely in deep space was something we could do, and Lunar Orbiter continued that tradition. That's one of the reasons, as an aside, why I say, "I think we're ready to go to Mars now, because we've learned all those lessons." Indeed, we probably have more information about the surface of Mars, its properties, and what we need to do to work in the vicinity of Mars, through its atmosphere, and on its surface, now, today, than we had even before Apollo 11 landed on the Moon.

Precursors are nice; they will provide and can provide a great deal of interesting scientific information but, from an operational point of view, I am not one of those that thinks we should hang our hat on the success of precursors before we begin that process of extensive exploration and ultimate settlement of Mars.

What were the main criteria used in the selection of the Apollo lunar landing sites?

Every landing site was selected, in part, to give us a different look at the lunar materials, and that included their composition. I guess one would say there were two major criteria in the selection of landing sites. One was the desire to sample as broad a range of rock and chemical types of lunar materials as we could, and, secondly, to look at as broad a range of processes, of structural features, geographic features, if you will, as we possibly could. And, indeed, the sharpest contrasts would have existed compositionally, in reference to your question, between those landings on the lunar maria—the dark parts of the Moon, the mare basalt areas—the highlands, such as in Descartes at the Appenine front, in the south and north massifs of later missions; and, of course, the landing at the Apollo 14 site, known as Frau Mauro, where we ran into this material I referred to earlier as KREEP. So, there were three major contrasting compositional regions sampled, and within even the individual mare sites, there were very, very great differences. For example, the Apollo 11 and 17 sites exposed us to very titanium-rich basalts;

whereas, the Apollo 12 and Apollo 15 sites were significantly lower in that element. That's interesting in regard to Helium 3, because it turns out that Helium 3 tends to be concentrated in the higher titanium soils. Helium 3 comes from the Sun, along with these other gaseous elements, and is absorbed in the lunar soils from the solar wind. But even at a given site, and I looked probably most closely at the Apollo 12 sites, you get contrasts. Some flows will be titanium poor, some flows titanium rich, and so the heterogeneity of both lunar composition and lunar processes is very great.

George M. Low at NASA's Lewis Research Center. (NASA Photo 70-26105.)

CHAPTER 13

GEORGE M. LOW

(1 9 2 6 – 1 9 8 4)

George Michael Low was born George Wilhelm Low on June 10, 1926, near Vienna, Austria. His parents, Artur and Gertrude Burger Low, operated a large agricultural enterprise in Austria. After the German occupation of Austria in 1938, his family emigrated to the United States in 1940. In 1943, Low graduated from Forest Hills High School, Forest Hills, New York, and entered Rensselaer Polytechnic Institute (RPI). His education was interrupted between 1944 and 1946 while he served in the U.S. Army during WWII. During this period, he became a naturalized American citizen, and legally changed his name to George Michael Low.

After military service, Low returned to RPI and received his Bachelor of Aeronautical Engineering degree in 1948. He then worked at General Dynamics (Convair) in Fort Worth, Texas, as a mathematician in an aerodynamics group. Low returned to RPI late in 1948 and received his Master of Science degree in Aeronautical Engineering in 1950. In 1949, he married Mary Ruth McNamara of Troy, New York. Between 1952 and 1963, they had five children: Mark S., Diane E., George David, John M., and Nancy A.

After completing his M.S. degree, Low joined the National Advisory Committee for Aeronautics (NACA) in 1949, as an engineer, at the

Lewis Flight Propulsion Laboratory in Cleveland, Ohio (now the John H. Glenn Research Center at Lewis Field). He became head of the Fluid Mechanics Section (1954-1956) and Chief of the Special Projects Branch (1956-1958). Low specialized in experimental and theoretical research in the fields of heat transfer, boundary layer flows, and internal aerodynamics. In addition, he worked on such space technology problems as orbit calculations, reentry paths, and space rendezvous techniques.

During the summer and autumn of 1958, preceding the formation of NASA, Low worked on a planning team to organize the new aerospace agency. Soon after NASA's formal organization in October 1958, Low transferred to the agency's headquarters in Washington, D.C., where he served as Assistant Director for Manned Space Flight Programs. In this capacity, he was closely involved in the planning of Projects Mercury, Gemini, and Apollo. In October 1960, Low recommended that a preliminary program for manned lunar landings be formulated to provide justification for Apollo and to place the schedules and technical plans on a firmer foundation. This proposal led to his being selected as chairman of a committee that conducted studies leading to the Apollo lunar landing program, the facts of which helped President Kennedy in 1961 to declare his goal of landing a man on the moon before the end of the decade. That same year Low became deputy associate administrator for Manned Space Flight where he was responsible to the associate administrator for Manned Space Flight in the management of the Gemini and Apollo programs and the field centers directly associated with those programs.

In February 1964, Low transferred to NASA's Manned Spacecraft Center in Houston, Texas (now the Johnson Space Center), and served as deputy center director. In April 1967, following the Apollo 204 fire, he was named manager of the Apollo Spacecraft Program Office (ASPO), where he was responsible for directing the numerous redesign changes made to the Apollo spacecraft in order to get the program back on track. Low is also credited within NASA for conceiving the idea of the Apollo 8 lunar orbital flight and proving its technical soundness to others.

George Low became NASA deputy administrator in December 1969, serving with Administrators Thomas O. Paine and James C. Fletcher. As such, he became one of the leading figures in the early development of the Space Shuttle, the Skylab program, and the Apollo-Soyuz

Test Project. From September 1970 to May 1971, he served as acting administrator of NASA.

Low received numerous awards and honors during his distinguished career including the NASA Distinguished Service Medal (twice awarded), NASA Outstanding Leadership Award, Honorary Doctor of Engineering from RPI, and Honorary Doctor of Science from the University of Florida.

Low retired from NASA in 1976 to become president of RPI, a position he held until his death on July 17, 1984.

Editor's note: The following are edited excerpts from two separate interviews conducted with George M. Low. Interview #1 was conducted while Low was deputy director of NASA's Manned Spacecraft Center (now the Johnson Space Center) in Houston, Texas, on May 1, 1964, by Dr. Eugene M. Emme, NASA historian, assisted by Mr. Jay Holmes of the Office of Manned Space Flight and Mr. Addison Rothrock, NASA consultant. Interview #2 was conducted by Robert Sherrod on January 16, 1974, at NASA Headquarters while Low was serving as deputy administrator of NASA.

Interview #1

I believe we should start our historical review of NASA's manned space flight program during the administration of President Kennedy by asking you the places and events that you had personal contact with the late President.

Unfortunately, I had only one contact with President Kennedy, and this was six days before he was assassinated. As you will recall, he visited Cape Canaveral on November 16, 1963, for a fairly brief tour. He reviewed the manned space flight program, flew over the new Cape Canaveral area—the so-called Merritt Island Launch Area (MILA)—in a helicopter, and went out to sea to watch a Polaris launching from the submarine U.S.S. Andrew Jackson.

My own part in the proceedings that day was a relatively short one. The President arrived during the morning of the 16th on the skid strip at the Cape and immediately went in a motorcade to Launch Complex 37, the Saturn launch complex. Outside of this complex we had the first Gemini spacecraft, the spacecraft that was subsequently launched on April 8, 1964.

The President's first stop on the tour was at this spacecraft, where we briefed him on the Gemini program. During this briefing, which was conducted by Astronauts Gordon Cooper, Gus Grissom, and me, we discussed the Gemini program and its current status.

I was very impressed, at that time, by the President. As I mentioned, it was the first time I had met him personally. I was particularly impressed by the questions he asked and by his detailed knowledge of the program. He had been in St. Louis, where the Gemini spacecraft was being produced, earlier in the year on September 12, 1962. At

310

that time he had been exposed to the details of the Gemini program. He recalled his visit to St. Louis and having seen the same Gemini spacecraft while he was there. He asked a number of very penetrating questions about the program and then proceeded into the blockhouse for a briefing by Dr. George Mueller [associate administrator for Manned Space Flight] on the Saturn program.

Historically, looking at the Kennedy administration and the national space program, particularly the manned space flight aspects of the national program, you had a key role in NASA's side of the work that went on, and in the historic decision announced by the President before the Congress on May 25, 1961. The decision to land an American on the moon in this decade followed a rather torturous history as far as manned space flight was concerned. It started quite apart from the hard work and problems of the Mercury program, which was coming into the launch phase of its program in early '61. President-elect Kennedy had appointed a committee, under Dr. Jerome Wiesner,[1] to make recommendations to him concerning the national space program. This, in turn, followed a rather crimping budgetary treatment of Mercury and Apollo in the last Eisenhower budget.

I wonder if we might talk now about this transition period from the Eisenhower administration to the Kennedy administration and how this affected you people in the NASA manned space program. Obviously, the Wiesner report of January 10, 1961, was a detailed written report, an unclassified version of which was public knowledge two days later. It was very critical of Mercury and the NASA leadership. What about this period of transition from one administration to the other?

Before I go into details on that, I think it might be interesting to note that three years ago today, early in May 1961, four or five days before Alan Shepard's flight, most of us concerned with the manned space flight were very deeply involved in the Mercury program. We worried about the day-to-day details of that program, and our future was tied completely to the Mercury program. But we did not know whether there would be any manned space flight program

The first Apollo 11 sample return container, containing lunar surface material, arrives at Ellington Air Force Base by air from the Pacific recovery area on July 25, 1969. Happily posing for photographs with the rock box are (left to right) George M. Low, manager, Apollo Spacecraft Program, Manned Spacecraft Center (MSC); U.S. Air Force Lt. Gen. Samuel C. Phillips, Apollo program director, Office of Manned Space Flight, NASA Headquarters; George S. Trimble, MSC deputy director (almost obscured); Eugene G. Edmonds, MSC Photographic Technology Laboratory; Richard S. Johnston (in back), special assistant to the MSC director; Dr. Thomas O. Paine, NASA administrator; and Dr. Robert R. Gilruth, MSC director. (NASA Photo S69-39984.)

beyond Mercury. We did not know whether the country would support a major effort beyond Mercury in manned space flight.

Two years before, ground had not yet been broken on the Manned Spacecraft Center at Houston. As you know, there was concern there, with 2,500 people working on a day-by-day basis on the Gemini and Apollo programs. In the same time period, launch vehicle development facilities were being built in Alabama [The Marshall Space Flight Center], Mississippi [Mississippi Test Facility now Stennis Space Center at Bay St. Louis] and Louisiana

[Michoud Assembly Facility at New Orleans]. There were launch facilities in Florida and a team of contractors to build boosters and spacecraft for the Gemini and Apollo missions started work.

This, perhaps, paints a clear picture of what happened in a three-year period. The work leading up to that, of course, took place during 1960 and early 1961. The Space Task Group at Langley Field, the group that became the Manned Spacecraft Center at Houston, prepared preliminary plans for a program beyond Mercury, early in 1960. I don't recall the exact dates, but I believe that in the spring of 1960 the team from the Space Task Group went out and briefed all of the other NASA centers on our thoughts for manned space flight beyond Mercury.

This effort culminated in a briefing to NASA Administrator [T. Keith] Glennan on July 9, 1960. At that time we gave Dr. Glennan, Dr. Dryden, and Mr. Horner, who was then associate administrator of NASA, a description of the advanced manned space flight program as we then saw it, its scheduling implications, and its costs.

Now at that time, the program we had planned was one not leading to a lunar landing but one that would stop after the circumlunar flight.

Gemini was not in it?

Gemini was not in the picture. I was trying to recall some of the planning that occurred prior to this, in 1959 or early 1960. This was done by the Goett Committee[2] . . . The Goett Committee included people like Al Eggers from Ames and Max Faget from Space Task Group. I served on it representing Abe Silverstein. There were several other people—I don't recall who they were. As I recall, the Committee deliberated whether the next major step in manned space flight should concern itself with a large space station program or a deep space program. The basic conclusion of the [Goett] Committee was that a deep space program (and by deep space I mean anything beyond earth-orbital activities) would lead to a quicker, focused advancement of the technology needed for manned space flights.

On the other hand, the [Goett] Committee also felt that making a lunar landing, or planning for a lunar landing, was a step which was too far beyond what we were willing to talk about at that time. So the recommendation of the Committee was that the manned space flight effort of NASA should be focused on a circumlunar flight.

Out of this Committee's recommendations, the Space Task Group prepared a program plan. They presented this plan, as I mentioned before, to the NASA Research Centers, with primary emphasis on informing the centers of the kinds of research which should be carried on to make such a program possible.

Who was Chairman of the Space Task Group [STG] team?

Bob Piland was Chairman of the STG team. Piland, at the time, was deputy to Max Faget, who was chief of one of three divisions in STG.

Were industry studies emanated?

Industry studies did not come until later. I have here copies of some of the slides that were used in this briefing to the centers. This effort culminated in the presentation I mentioned earlier, to Dr. Glennan, Dr. Dryden, and Mr. Horner on July 9, 1960. At that time Dr. Glennan gave preliminary approval to the program. The briefing was also conducted in the context of a forthcoming NASA-Industry Conference and the material which I was to present.

The general subject of the NASA-Industry Conference was to describe industry studies we would soon request for a circumlunar flight program, but at the same time to inform industry that we had no approval for the program beyond these studies; that we were, at this time, only talking about a study effort . . .

At the meeting on July 9, the name Apollo was approved for the program. I had already prepared my paper for the NASA-Industry Conference before the meeting on the 9th, and later added a comment, "and we will call this program 'Apollo.'"

Now remember, at this time we still were talking about only circumlunar flight. In fact, we said Apollo was a program with two avenues of approach—one of them, the main stream, being the circumlunar flight program. We had planned a program that, within this decade, would lead to a circumlunar flight; beyond this decade,

Apollo would eventually lead to a lunar landing and to planetary exploration. We also proposed that, within this decade, Apollo could lead to, or could be part of, the space station program . . .

At the time we had people at Space Task Group considering what we could do beyond Mercury. I don't think we were seriously thinking about Mercury Mark II, which later became the Gemini program, until the summer of 1961. The Gemini program was approved in December 1961. Also, you must remember that most of us at the time were spending full time on Mercury, and very little time was available to consider what would happen beyond Mercury. I mentioned that Bob Piland was the sparkplug for Apollo at the Space Task Group, and I doubt that he had more than two or three people working with him on future programs out of the total people assigned to the Space Task Group.

In Washington, we had a very small group in Manned Space Flight. The only one in Washington who really spent any time on advanced programs was John Disher, and he probably didn't spend more than ten percent of his time on that because he was also involved with the Mercury program. So this was strictly an extracurricular activity which we fit into whatever time we could spare from the Mercury program. This was true of Bob Gilruth and his top people and of all the people here in Washington, I'm sure.

That brings us, roughly, through September 1960. Were the future programs or goals of NASA discussed at the Williamsburg Conference?

I was not at the Williamsburg Conference . . . I believe others can speak on that much better than I can. While the studies for the circumlunar flight were going on, we became concerned again as to whether we were going far enough with the circumlunar flight or whether we should really focus our attention on a lunar landing.

I had forgotten about a memo which my secretary [Mrs. Lillian Stutz] dug out for me this morning, which I will read to you. This is a memo dated October 17, 1960, for the director of Space Flight Programs, Dr. Abe Silverstein, on the subject of the manned lunar landing program.

Paragraph 1 states, "It has become increasingly apparent that a preliminary program for manned lunar landings should be formulated. This is necessary in order to provide a proper justification for Apollo, and to place Apollo schedules and technical plans on a firmer foundation."

The memo went on to say in paragraph 2, "In order to prepare such a program, I have formed a small working group consisting of Eldon Hall, Oran Nicks, John Disher, and myself. This group will endeavor to establish ground rules for manned lunar landing missions, to determine reasonable spacecraft weights, to specify launch vehicle requirements, and to prepare an integrated development plan including the spacecraft, lunar landing and take-off systems, and launch vehicles. This plan should include a time phasing and funding picture and should identify areas requiring early studies by field organizations.

Paragraph 3 . . . "At the completion of this work we plan to brief you and General Ostrander on the results. No action on your part is required at this time. Hall will inform General Ostrander that he is participating in the study." Signed by George M. Low, Program Chief, Manned Space Flight. And there is a notation under it in pencil, "Low, O.K.," signed "Abe."

What was your motivation for this particular memo? What triggered it?

I knew you would ask that question, and I don't know.

It was just a sense of timing?

This was the time, of course, that we were beginning to discuss with industry what the Apollo program was. We were also quite concerned, of course, that in the subsequent year's budget, which was being prepared at that time, there were insufficient funds for any major lunar program. And we felt it would be most important to have something in the files, to be prepared to move out with a bigger program should there be a sudden change of heart within Government, within the administration, as to what should happen. This memo (I was just looking at the date) was written during the Eisenhower administration and before Election Day.

We really have two main historical paths from here on out. One is the President's Scientific Advisory Committee (PSAC) and its role with regard to Mercury, which becomes more prominent with the new administration, and the other is the whole process of the lunar landing decision and how that came about in May. Do you think we can develop these two themes chronologically, or should we take one story and then do the other?

It would be easiest for me to discuss what happened within NASA in the lunar program and then go back to discuss what happened with Mercury at the same time, because PSAC was involved in that.

Now, one reaction before I forget. The people we talked to have said there was a general uncertainty existing in NASA in the December-January-February period.

Yes, and I think we all felt this quite keenly. As you know, the last Eisenhower budget carried a paragraph which said we'll complete Mercury and then future studies will be needed before we could decide what else we could do in manned space flight.

Was Dr. Glennan the only one carrying to the Bureau of Budget and the White House the thought that had gone into future manned missions? How was the Eisenhower budget determined? Did Glennan give you the planning go-ahead right from the start in July?

In July, we had the planning go-ahead. I believe Dr. Glennan, Dr. Dryden, and Dr. Seamans, who had joined the organization in the summer or fall of 1960, carried the story forward to the Bureau of the Budget.

The next major step in the planning within NASA was essentially an outgrowth of the Space Task Group that I formed on October 17, 1960. We put together a preliminary story during November and

December, 1960, and on January 5, 1961, presented the program for manned lunar landing to the Space Exploration Council of NASA. This Council, I believe, consisted of the directors of the major centers and Dr. Glennan and his immediate staff in Headquarters.

This was a full day of briefings?

This was a full day of briefings, following general guidelines which were developed by this initial small task team I had. There were brief- ings by the Space Task Group; Mr. Faget made a presentation; and Dr. von Braun made a presentation for the Marshall Space Flight Center.

The presentations showed that there were major disagree- ments within NASA as to how the problem of a lunar landing should be approached. The people at the Space Task Group were primarily interested in the so-called direct approach of going to the moon while the Marshall group felt very strongly that an earth orbit rendezvous approach should be used.

At the same time, of course, John Houbolt of Langley Research Center was developing his lunar orbit rendezvous approach, but I don't think this was described in any detail at the January meeting . . .

It was something we were all aware of. We knew Houbolt was working on it, and occasionally people told us about it, but at first nobody thought it was a worthwhile approach. This, historically, was the case with everybody who looked at it. We were horrified at the lunar rendezvous approach the first time we saw it, and it was only after we studied it in depth, as Houbolt was doing at the time, that we became convinced that this was really the way to go.

Were you discussing the lunar orbital mission, or were you really focused on the lunar landing in this January 5 meeting?

In the January 5 meeting, the title of the presentation was, "A Program for Manned Lunar Landing," so it was an effort on January 5 to show top NASA management what could be done to extend the then-existing Apollo program to a manned lunar landing pro- gram. But it was sort of a diversified . . .

When you say "then-existing Apollo program," the Apollo program at that time was a proposal, a series of studies being done by industry, wasn't it? And that was what Eisenhower had cut out of the budget just the same month, wasn't it? Or maybe it was just before or after that?

It was a series of studies proposed to industry. I believe we were supposed to spend no more than a million dollars, and the goal of these studies was circumlunar flight. This is a point I can't emphasize too strongly, that the studies ended with circumlunar flight and not with a lunar landing.

It is very important to stress that this was a lunar landing program examination at the January 5th meeting as distinct from the Apollo studies, which were circumlunar studies. Had you gone over the Air Force studies, the contract studies on lunar landings, to any extent?

The people on my task force had gone over this. In fact, I think we assigned one of our people to review these studies. I don't remember them in any detail, however. One recollection I have of the meeting is that Dr. Glennan listened very politely, as he always did, but emphasized at the end of the meeting that we had absolutely no authority to go ahead with any program beyond the Mercury program, except for the studies for the Apollo circumlunar program; that he could not really authorize us to pursue these matters any further. He put a heavy damper on this entire effort. His function then, of course, was to carry out the wishes of the administration, which had been that there would be no major emphasis on manned space flight following the Mercury program.

The minutes of that meeting, which are very terse and general, do show that he placed that on the record. Nevertheless, did the Saturn program continue at that time?

Yes, but not beyond the Saturn I program. They were developing the second stage but weren't developing any advanced Saturn hardware. The C-2 was merely in a study phase. Since my main association at that time was with the spacecraft end of it, I probably slighted the launch vehicle, and I'm glad you mentioned this. I think one of the most important decisions in the whole lunar landing program was NASA's decision, and I probably should say Abe Silverstein's decision, to base the future launch vehicle programs on hydrogen technology for the upper stages. This, at the time, was a very daring decision.

What was that approximate date?

I would say 1959. You'll find that when the decision was made to have the second stage on the Saturn I, the second stage then was hydrogen. I think it happened in December 1959. It became public knowledge a short time later. When the budget was presented to Congress in early January 1960, they asked for a supplemental appropriation to get the J-2 engine program going. But remember, at that time there was no J-2 engine. The RL-10 engine was not working too well. [It has turned out to be an excellent engine now.] And we really knew very little about insulation and bulkheads and all of these problems. Yet the decision was made by a committee chaired by Silverstein to base the whole future launch vehicle program on hydrogen technology. It was really one of the most important decisions in the program, one of the key decisions that allowed us to go ahead with the Apollo lunar landing program in 1961.

Is there anything more as a consequence of this January 5 meeting?

As a result of this meeting, Dr. Seamans established the Manned Lunar Program Planning Group. This planning group met for the first time on January 9, 1961. Again I was chairman of that group. Members were Oran Nicks, E.O. Pearson, Al Mayo, Max Faget, Herman Koelle, and Eldon Hall. The purpose of this group was to prepare a position paper which would answer the question, "What is NASA's Manned Lunar Landing Program?" This group met almost full-time for a week or two and prepared a report which was presented in final form to Dr. Seamans on, I believe, February 7.

Pictured inside the Complex 34 firing room at the Kennedy Space Center during the Saturn 205 countdown demonstration test on September 16, 1968 are (left to right): Dr. Kurt Debus, director Kennedy Space Center; Dr. George Low, manager Apollo Spacecraft Program Office; and Rocco Petrone, director Launch Operations. (NASA Photo 107-KSC-368.)

In this group we had representatives who favored the earth-orbital rendezvous approach; we had others who favored the direct approach to the moon. Again, Houbolt was not represented. We did assign to one of our members, however, the task of studying and looking into Houbolt's approach and studying the lunar orbit rendezvous method.

He came back later and said he had studied it and his basic conclusion was that the lunar orbit rendezvous method would not work . . . It was Max Faget who did this study, and based on his preliminary look, it did not appear to be an appealing approach. Max, at the time, was strongly in favor of the direct approach to the moon. A year or so later, Max Faget himself became one of the strongest proponents for the lunar orbit rendezvous approach, so it's interesting to see that outstanding technical people do change their minds and their conclusions occasionally . . .

I think we were fortunate in a number of areas to have people in the agency with the foresight to start hardware in critical areas. The F-1 engine, the Saturn I booster, and the hydrogen decision were the three things that allowed us to jump in with both feet in May 1961. Without those we couldn't have done it.

I'll come back to the Mercury program. Gagarin's flight was on April 12, 1961. On April 12 there was sudden interest again, in this country, in manned space flight. On April 11, the day before Gagarin's flight, I was in the middle of a presentation to the House Committee on Science and Astronautics, which was chaired by Congressman Overton Brooks [Democrat from Louisiana] at the time, on the manned space flight picture and defense of the budget for manned space flight.

I was about half way through the presentation when it was time for lunch and Committee adjournment. I had with me a movie of Ham's flight.[3] Ham had flown in January, and this was a short sound film of the first flight using a chimpanzee in a manned space flight capsule that this country had made.

The Chairman said, "Let's start with the movie when you get back tomorrow." That night Gagarin flew. The next day I did not go back to complete my testimony. Mr. Webb, Dr. Dryden, and Dr. Seamans (I'm not sure whether Dr. Seamans was there as a witness or not) presented an overall picture of where we stood in manned space flight, what the Russians had done and what they could be expected to do, and what we had done.

It was probably one of Mr. Webb's first appearances before that Committee and he did an outstanding job in avoiding panic in the country by the way he presented the NASA picture on the day after the Gagarin flight. Incidentally, that hearing was not held in the normal Committee room. It was in the Caucus Room, and it was filled completely with interested bystanders.

The following day, we went back to complete my presentation of manned space flight. Dr. Seamans and I were before the Committee, back in the Committee room again. One of the things I remember was that we decided not to show the film. We thought it would not be in our best interest to show how we had flown a monkey on a suborbital flight when the Soviets had orbited Gagarin. The Chairman did say, "Well, we thought we were going to start with the movie." We looked around and the projectionist wasn't there, and we fumbled and said, "We don't have it with us today."

But we did go on, and during the hearings, Representative [David S.] King of Utah noted that the Soviets were being quoted as saying they would land on the moon in 1967. He asked whether we could do it. The only background that Seamans really had at that time was the February 7 report, and under pressure he said the goal might well be achievable. This is where the 1967 date first appeared.

King gave him the 1967 date and said that was the 50th Anniversary of the Bolshevik Revolution . . . can you make it? That's the way the words were put into Seamans' mouth.

Incidentally, I'm just checking the February 7 report. it showed manned flights to the moon in the 1968-1970 time period.

Following this, then, there are a number of items in your history where Vice President Johnson asked the agency for a plan for manned space flight. The Fleming Committee was organized in about this time period. The purpose of the Fleming Committee was to get better schedule information, better cost information, following up on the February 7 report with a much larger effort to come up with more specific details of what the program should cost, how it should be done, and where it should be done.

I think it's interesting that the men on your Committee were members of the Fleming Committee.

That's right. The same people participated and carried on with the work they had done. One way of looking at it is that it went from the small Space Task Group, which I formed on October 17 [1960], to the Low Committee of January 7 [1961], to the Fleming Committee formed in April—one flowing into the other and each one giving more specific details.

And you still had not flown the first Mercury astronaut?

We still had not flown Shepard's flight. Of course, it was immediately after Shepard's flight that these things were available. The Fleming Report was by no means complete.

Webb was briefed on the results around the third week in May.

But at the same time, NASA management went forward and presented the plan to the administration. This was between Shepard's May 5 flight, as I recall, and President Kennedy's speech on May 25, which gave the country, and all of us who were working on the program, the go-ahead and boost we needed.

That's a very important historical chapter we have just reviewed. It's the internal NASA work contributing to the lunar landing decision of May 1961. Does that cover it fairly well so we can go back now to the Mercury program and the transition from one administration to the other? What was the impact of the Wiesner report upon the Mercury program? That might be a way to get this kicked off . . . and then, of course, you'll want to bring out some of the problems that the program was involved with, the booster, etc.

First of all, I think I should say that we did have complete and full support during the Eisenhower Administration and under Glennan for the Mercury program. The thing I may have questioned before was the continuation of future manned space flight beyond Mercury. But there was no time in the Mercury program where we were held back because of funding, or because people told us to go slowly. In fact, the pressure was always in the other direction. It was, let's get on with Mercury.

I remember Dr. Glennan, on many occasions, driving us very hard and, of course, this feeling permeated throughout the entire Mercury program. So the impact of the change in administration on the Mercury program was a minor one in that sense, if any.

In the late fall or early winter of 1960, the Mercury program was approaching the manned space flight stage. We'd had unmanned flights on the Redstone and on the Atlas; we'd had successes, and we'd had failures. But the program was still moving along as quickly, or more quickly, than any other program had ever moved. We drew comparison curves, in connection with Gemini and Apollo. Mercury was moving exceedingly well.

The first Redstone flight with a Mercury capsule was on December 19, 1960.[4] There had been an unsuccessful preliminary attempt at this flight in November. We then went on with the Redstone flight in December, and the Ham flight was in January.

At about this time period also, and I don't know whether it was January or February, a number of us felt that before very long we would be ready to tell our management that we were ready to make a manned space flight, the suborbital flight on the Redstone. The flight was scheduled for about April. We felt that the administrator would have to go forward to the president and say, "We're now ready to make a manned flight." Bob Gilruth believed that if the new president were faced with this decision, not knowing the agency and not knowing the people in the agency at that time, he would be hard pressed to say, "Go ahead." He would want to ask questions: Do these guys know what they are doing? Are they really ready? Or do they just want to go ahead?

So we suggested to Dr. Dryden at the time, and this must have been in the interim before Mr. Webb came on board, that we should have a committee appointed by the president to check on the Mercury program. Dr. Dryden went to see Jerry Wiesner, who had already been appointed the President's Science Advisor, with this idea, and together they appointed a committee which was not really a President's Science Advisory Committee [PSAC], although it had some PSAC membership on it. It was chaired by Don Hornig, who is now the President's science advisor and chair of one or two panels.

It was called a Subcommittee on Manned Space Flight?

It was a special subcommittee, but it included many members who normally were not on the President's Science Advisory Committee. This Committee met in March and traveled around the country to McDonnell, to the Cape, to Langley, and to NASA Headquarters in examining the Mercury program. There were some very competent people on this committee, some who had been involved in this kind of business before, and others who had not.

I think to me the most gratifying experience was to see a man like Don Hornig, who had contributed to the Wiesner report, come in with a very open mind about the program. He was an excellent chairman but also one with a fairly negative approach to manned space flight as a whole, at least initially. To see him argue, on the last day of that week, with Jerry Wiesner, and support the benefits of manned space flight, saying what a wonderful program Mercury was, was very gratifying.

You reported to the NASA senior staff meeting on March 15 that the Hornig Committee, having visited Langley, St. Louis, and AMR [Atlantic Missile Range], "had a better feeling now for the system as it is."

The Hornig Committee was the major influence of the Wiesner group on the Mercury program. Now we were at perhaps the most difficult time in the whole Mercury program. I don't have a flight schedule in front of me now, but in a period of, I think, ten days we got off three or four launches, each one of which had to be successful before we could go on with Shepard's flight.

So we were in a situation where, first of all, there was major Soviet activity leading up to Gagarin's flight. We had major difficulties with the Atlas between the MA-1 flight and the MA-2 flight, and NASA as an organization stuck its neck out with the MA-2 flight with a temporary fix to do that. We had had two Little Joe failures, and we needed a successful Little Joe flight. We had to get MR-2 off. We had landing bag difficulties with MR-2 and had to fix that. All of these things had to happen before we could fly Shepard. On top of that, we had to support Hornig Committee investigations.

How about the Russian dog flights?

One point I did not mention before is that, although the initial Hornig report appeared to be very good, some of the medical members of the Hornig Committee caucused the night before the report was made, in Washington, and turned in a very, very negative report. They recommended that we could not possibly take the risk of flying a man in a weightless flight, where he would be weightless

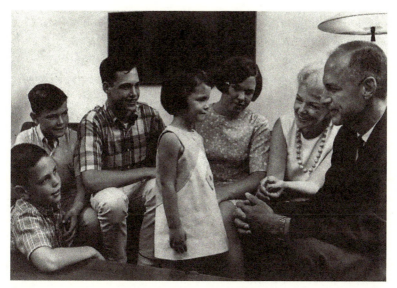

George Low is shown with his wife and children in this photo taken on February 3, 1970. Pictured (left to right) are John, G. David, Mark, Nancy, Diane, Mary, and George. G. David Low became an astronaut in 1985 and flew three Shuttle missions (STS-32, STS-43, and STS-57), and logged over 714 hours in space, including nearly 6 hours on a spacewalk. (NASA Photo S70-26140.)

for five minutes, without many additional tests of monkeys. In fact, they wanted us to fly 50 monkeys. We learned there weren't this many trained monkeys available in the country.

Was Bob Gilruth going to transfer the program to Africa, where they have plenty of chimps . . .?

They recommended additional space flights. It was an exceedingly difficult period to try to make all of the hardware work, while at the same time satisfing a group of uninformed people as to whether or not we should go ahead . . .

But what really disappointed—I think that is the word to use— many of us who were trying to get on with the program was that this was a group of people who, we were convinced, were not really informed on what they were trying to do. They were not approaching it like Don Hornig had, in a positive manner, and they caused an awful lot of difficulty at a time when we had enough difficulties. This really was the most difficult time in the Mercury program.

Until April 12 were you still really working to beat the Russians in a manned flight?

One of the deepest disappointments to me was when I got a phone call, the night of April 12, at 2 am, that Gagarin was up.

Is there anything in particular that should be put on the record with regard to the Redstone and those various problems that wouldn't already be on the record? Did you have more confidence, perhaps, in the Redstone than the Marshall people did? Wasn't it the MR-2 flight that had not followed the exact parameters precisely and yet was within the tolerances?

Very definitely at that time period there was a conflict between the booster people and the spacecraft people on whether or not an additional booster flight would be needed, and the booster people felt that they could not commit a man to the next flight. We did fly an additional Redstone flight without a spacecraft, and it was successful.

Did you have any comments to make on the post-Shepard flight and the apparent impact that the Shepard flight might have had on the president and the American people?

I recently made a comment on this subject in a paper I gave on October 1, 1963, the fifth anniversary of NASA. I just happened to be giving a talk at a technical meeting on the West Coast—a luncheon speech—and I had a comment in there about how surprised we were at the tremendous interest that the people and the press had in Shepard's flight and all subsequent flights. In a press conference following this speech, one of the reporters asked, "How naive could you have been to think that there wouldn't be this interest?" Yet I still don't think we were any more naive than the newspaper people themselves at that time . . .

I was tremendously elated the day after Shepard's flight. I remember coming back to town that evening. I got into the office just before quitting time (you know, it was a short flight then), and I invited everybody and anybody that I could find to a party at my home that evening. And my wife didn't know I was in town yet. [Laughter] Then I stopped at the liquor store on my way home. It was probably one of the best parties we ever had—in Washington, at least, following any flight.

And I recall also that John Disher and Warren North and I went almost directly from the party to my office, the next morning, and we decided, "Let's put some more finishing touches on what we can do on the lunar landing program." Now what I did not know at the time was that Abe Silverstein was meeting with Webb, Dryden, and Seamans, and others, and actually spent part of that day with McNamara, being about ten steps ahead of what I was trying to do in my office. So this is the kind of impact that this made on everybody.

The culmination of this effort, of course, was President Kennedy's address to Congress on May 25, 1961, committing this nation to the goal of a manned lunar landing.

Could you discuss some of the steps taken after the decision to land an American on the moon was announced by President Kennedy on May 25? Was not the lunar landing mode the next major decision?

The launch vehicle people formulated detailed plans for the facilities they would need and for the configuration of the launch vehicles which would be needed.

On the spacecraft end, during the summer of 1961, we prepared specifications and requests for proposals for the Apollo spacecraft. These were mailed out during the summer. And this, of course, led to the selection of North American for the contract for the Command and Service modules.

Also during this time period, people here, and particularly at the Space Task Group, were doing more detailed designs of the lunar landing system, and—remember—this was a Lunar Landing

Stage, at the time, to go behind the Command and Service modules. We also started getting more and more interested in the lunar orbit rendezvous approach.

I'm trying to think of the timing here, and I believe John Disher could help you much more than I can about the detailed effort during this period of time. Brainerd Holmes came on board in October 1961. We discussed with him during the rest of the year various designs for the lunar landing stage, for the "direct approach" to the lunar landing. We were not really talking to him yet about the possibility of a different method or approach, because none of us had yet become convinced that the lunar orbit rendezvous method was the way to go.

I remember taking Joe Shea, who had come aboard in December 1961, to introduce him to the Space Task Group. We spent a good part of the day being briefed, by the Space Task Group and people from Langley Research Center, on the lunar orbit rendezvous approach. It was about January 1962 that Dr. Gilruth and his key people became convinced that lunar orbit rendezvous was the way to go.

Gemini was coming along at about the same time?

Gemini was coming along as an almost separate program at about the same time, with studies during the summer of 1961, the development plan prepared about November 1961, and the program being approved in December 1961.

And the booster studies were under way in the Golovin Committee during the summer of 1961?

That's right. It wasn't until December 1961, when we had the first meeting of Brainerd Holmes' Management Council, that we decided that the Saturn V should have five engines in the first and second stages. Until that time we had four engines.

I remember also, in this December meeting, Bob Gilruth saying, "Well, even if we don't go ahead with the approach we presently have planned (namely, the direct approach), we could use that extra booster power with any other approach that we would take." So he went along with it on that basis.

330

Gilruth had become convinced in favor of lunar orbit rendezvous?

He was beginning to be, in the December 1961–January 1962 time period.

So he had LOR in mind when he was supporting the five-engine advanced Saturn?

He definitely did. I recall, at the first meeting of the Management Council, Gilruth said he was not convinced we had the proper approach to the moon, but no matter which approach, he would like to see that fifth engine in there.

Was the major pacing item the boosters? You recall that in 1961 you had the decisions on the selection of AMR [Atlantic Missile Range, now named the Kennedy Space Center] as the launching facility; you had the Michoud and Mississippi static test area. These decisions were all announced late in December 1961.

During 1961, all of the major decisions in the program were made, except the LOR decision. We committed the land and the facilities. We committed to the contractors for, I think, all of our stages except for the lunar landing or LEM [Lunar Excursion Module] stage. So the only hole that was left in the program by the end of 1961 was the way to go to the moon, and that particular stage and its modules.

I didn't appreciate until now that our first Saturn launch on October 27, 1961, preceded the first orbital flight of Glenn (February 20, 1962). I think you might want to put on the record that we really wanted to try to orbit a Mercury astronaut in the calendar year 1961, as against doing it the same year as the Russians had done it. Was there any thought like that in NASA Headquarters?

From our point of view, and I speak of both the Space Task Group and my office at NASA Headquarters, we were always very schedule conscious in Mercury, and we had set ourselves a goal of getting a manned flight off in 1961. We had made these commitments and we just wanted to maintain and keep these commitments.

Unless you establish a goal for yourself in any program, whether it's going to the moon or orbiting a man, you will never make any schedules. So once we had established these goals, and once we had told the public and our management that this was what we were going to do, we were making every effort to do this. I think it was more that kind of a personal commitment we had all made as opposed to getting it off in the same year as the Russians did.

As you know, we came very close to getting Glenn's flight off in 1961. It slipped into January 1962, and then we had the weather problems in January that put it into February. But, again, I don't think I ought to apologize for any of the wonderful people working on Mercury for not making it until February, because it was still a tight schedule all the way through.

I guess the other major area we would like to cover before we conclude would be the mode decision of 1962. At what point did the arguments begin to break down and coalesce into pretty much a unanimous decision?

There were arguments at every step of the way. As I said, there was a time when Max Faget, early in 1961, was very much against this. By late 1961, he was convinced this was the only way to go, and so was Bob Gilruth. Coming into Washington after I had spent some time at Langley, I was probably starting to support and believe in this kind of a decision early in 1962.

At that time, we had organized our Systems Engineering organization under Dr. Shea, whose task it was to decide what mode to take to the moon, or how to go to the Moon, and how to synthesize the entire program. I don't know specifically when Joe and his people became convinced that LOR was the way to go.

How about the Huntsville people?

They were probably next in line. I do recall Joe Shea becoming convinced fairly early and then setting out to study this mode in real depth, and Brainerd Holmes beginning to lean toward LOR in the spring of 1962. And then Brainerd, through the Management Council, and through a series of technical briefings from Joe Shea's people, convinced the Huntsville people that LOR was the way to go. In the end, the Management Council unanimously recommended the lunar orbit rendezvous approach. Shortly thereafter, the recommendation was presented to general management by the Management Council. The next step was to convince Jerry Wiesner's people that this was the way to go. It took all of that summer to get everybody convinced that this was the approach we should take. In the meantime, of course, we had prepared our contractual work statements so that we could move very quickly with a contractor selection in this area.

We'd had industry studies, we'd had in-house studies, and one by one, as people looked into this more and more thoroughly, they became convinced as we had that LOR was the proper approach.

It's been said, and I'm not sure it's true, that Houbolt almost had to publish the LOR concept in *Astronautics and Aeronautics*. (I think it was February or March 1962 that it appeared) to really begin to catch thinking on the lunar orbital mode. Is this true?

Well, Houbolt did write a number of letters to Seamans and to Brainerd Holmes. In fact, as I recall, one of the first days that Brainerd Holmes was on board, he was handed a very long letter that Houbolt had written to Seamans saying that we really didn't know what we were doing. John Houbolt deserves a lot of credit for being as persistent as he was in pushing this forward. I'm now convinced that he was right.

In your office, were you becoming increasingly involved with the Apollo lunar program and Gemini? What was the relative concern you still had with Mercury in 1962, or was that largely under the Space Task Group?

It became less and less of a concern, of course. Perhaps I can best describe this in terms of organization. I had a very small office at

Headquarters initially, and by the summer of 1961 I had about ten people. Now this did not include launch vehicles. As you recall, there was a separate arm, reporting directly to the associate administrator at the time, for launch vehicles. But it did include the whole Mercury program, Gemini, and the Apollo spacecraft, as well as the mission planning for Mercury, Gemini, and Apollo. Out of those 10 people, which included three or four secretaries, only one, John Disher, was spending almost full-time on future programs.

After Brainerd Holmes came on board, my responsibility became that of directing the spacecraft and the flight missions. I organized four separate elements, one of which was entirely for Apollo, under John Disher, and this became the largest element of the group. Other elements were on Mercury and Gemini, under Colonel Dan McKee, and he might have had two or three men on Mercury, but the rest of the people were working on Gemini. There was an Operations group, under Captain Van Ness, which was mostly involved with Mercury in that period of time. So during the summer of 1961, the fall of 1961, and leading into 1962, more and more of my group was devoting attention to Gemini and Apollo, but we kept essentially the same number of people we always had on Mercury, which was only two or three people in Headquarters. I would say I probably spent less and less of my own time on Mercury.

Dr. Thomas Paine, right, administrator of NASA and George Low, Deputy NASA Administrator, monitor Apollo 13 prelaunch activities within the Launch Control Center at the Kennedy Space Center on April 11, 1970. (NASA Photo 70-H-506.)

Interview #2

Do you recall the story behind the adding of the fifth engine to the Saturn V?

My recollection is Brainerd Holmes' first management council meeting, which would have been in December of 1961. It was also the first time that I met Joe Shea. I think Joe was just coming onboard and sat in on that meeting—and we were in Brainerd Holmes' office in the building at 19th and Pennsylvania. We had a model of the Saturn V. I don't know what it was called at the time but it had four engines on it. Von Braun and a number of us were standing around the table before we went into the meeting, looking at this model, and von Braun started saying, "Here is where we want to put the other engine." That's the first I remember hearing about it. I asked him, "Why do you want another engine?" and he replied that, "Well, there's this big hole that is just

335

crying for another engine." He did, in a very forceful way, say that this is a capability that this vehicle should have and we ought to put it in.

I remember my job at the time was to look out for the Bob Gilruth half of OMSF, and Milt Rosen looked out for the von Braun half at the time. I remember talking to Bob Gilruth on the side and asking him what he thought of the added complication of one more engine. I was probably negative to that kind of change, but I'm not even sure of that. Bob said, "Look, I'm so dedicated to LOR and if that will be the only way to go for that kind of approach I'm sure we'll be able to use the added weight-lifting capability if we have the fifth engine. So let's go ahead with it." I think even at that time—long before the LOR decision was final— Bob was convinced that this was the only way to go.

What about Max Faget and the LOR decision?

That goes back to the Low Committee days. And in that context, we assigned different things to different people—a small group or a small committee. I assigned Max [Faget] to go to Langley to talk to Houbolt and to come back with an assessment of LOR to see whether we should consider it further. Max came back and spent quite a bit of time at the blackboard proving that we should not consider it. And none of us were smart enough to question him on it. It was a mistake on our part, on my part, I guess, not to probe further and turn Max around. The course of events might have gone differently. Max turned around not too long thereafter, and I think Bob Gilruth probably helped turn him around . . .

Have I ever talked to you about Jim Chamberlain and LOR?

No, I don't believe you ever have.

Jim Chamberlain had grandiose ideas about Gemini. He wanted to skip Apollo and take Gemini to the Moon. He came to meet with Abe Silverstein and me sometime in late 1960 or early 1961, and sketched out how we could use Gemini to land on the Moon using LOR.

336

A 1966 photo taken in Mission Control at Houston shows (left to right) Chris Kraft, Bob Thompson, George Low, and Bill Schneider. (NASA Photo S-66-33835.)

Regarding the crew selection process, do you know how that worked within the Astronaut Office?

On Mercury alone, I think Bob [Gilruth] started out with peer ratings. He asked each of the seven, "if you were unable to fly, assuming you wanted to fly first but were unable to, who would you like to see fly next? And why?" He used that information, and he asked Charlie Donlan and Walt Williams, both of whom were his deputies, for their input. And finally, based on all of that, he made his decision.

Gemini, Shea says, "was controlled completely by Slayton. Gilruth merely rubber-stamped." I think rubber-stamping is going too far. In the Apollo selections, when I was deputy director of the center [the Johnson Space Center], and even later when I was Apollo spacecraft program manager, Deke Slayton would come in with an overall plan and a detailed selection. He would say, "Here are the men I have that can fly; these are potential commanders;

337

so-and-so does not look like a commander to me but he would be a good back-up man; this man is very good at EVA, this man is particularly good at systems." He flew Gus Grissom first in Gemini and then again in Apollo because he thought so very highly of him, his engineering-pilot capabilities. He put all of these things together. And then Deke and Bob [Gilruth] and I would discuss Deke's plan at great length. Deke would discuss the overall plan and then say "now the guys I want to fly on the next selection—which probably was not the next mission—would be these and these would be the backup." The specific recommendations were based on the overall plan. I can't think of a single instance where Deke changed his mind based on our questions, or we changed Deke's mind. So if Shea wants to call it a rubber stamp, he may. I don't know of an instance where we disagreed with Deke's recommendations. But I would rather pay tribute to Deke's abilities in what he did than to say that no matter what Deke would have come in with, we would have accepted it.

I talked with Deke once about the only time he was overruled, and that wasn't by Gilruth, it was by Mueller, about putting Shepard on [Apollo] 14 instead of 13. The decision shows how lucky Alan Shepard was.

That's right.

He said he never knew why Mueller wanted that change made. Do you recall—

I was still in Houston. I remember Sam Phillips worrying a great deal about this one, and I remember Sam asking me on the side did I think Al should fly or not. I remember that Chris Kraft and I had many discussions about this. I think this was about the time of the Apollo 11 return . . . My concern about Al was whether a man who had taken command of a group of people and who really had not gotten his hands dirty for a while—had really been a leader of a team instead of a team member—would get back to work. Would he really knuckle down and take instructions from the flight controllers? Would he listen to them? Would he do what they asked him to, or would he throw his weight around? Both Chris [Kraft] and I were concerned about that. Chris and I, I think, both privately talked to Al about it. I know we talked to Gilruth and Deke about it and convinced them and ourselves that Al really wanted to fly so badly that he would knuckle down and he'd probably be one of the most dedicated pilots we'd ever had. And I think as it turned out, it proved that this was the case.

Now, why was Mueller against flying him on 13? Now I have to speculate. George did not like to lose a battle, to lose any kind of an argument, to lose any kind of a fight. It could be that by postponing Al's flying by one flight at least made him not lose completely. He sort of had to give in to the overwhelming support that Al Shepard had for flying, probably including Sam Phillips' advice, but finally said, "but will he really be ready? Remember, he hasn't flown since Mercury, he has not been in detailed training. How much time do you normally take for training? Don't you need that extra time period to train him?" And probably the other side, Deke and company, gave in to that argument . . .

I was one of two people who told Al Shepard that he would never fly again. Many years earlier, Bob Gilruth and I met in Bob's office with Shepard after we talked to Chuck Berry in detail. Chuck had told us that Al had Meniere's Disease, that this was incurable, and that Al would never fly—should not fly again. Al had high hopes always, and never gave up. Bob and I sat down with Al and told him that he would never fly again and might as well face it . . .

I remember Bob had a yellow couch and Al was sitting in it and we were sitting on the two sides, and it was one of the most difficult things I have ever participated in because telling a guy that what he's wanted all his life he won't be able to do is very, very difficult. Al, on his own, made himself flyable again. And I think from every point of view, from the point of view of compassion for the man, from the point of view of letting an incurable, whatever the incurable is, achieve his goal after he has been cured and what this can mean to people around the world who consider themselves incurable with more serious things, and knowing that Al would fly a good mission, I think there was just no question about it—that doesn't mean there weren't other people equally qualified or better qualified to fly—you put all of those things together and I think it was the right selection. I don't know what I'd be telling you if he'd blown it, but I don't think he would have.

Well, everybody likes to see a man come from behind like that. Especially one who was responsible for telling him about it.

Yes. And you know I didn't have all that much to do with selecting him at that time. I was off on the side making inputs so I can't take any credit for it. But I remember talking to Al and telling him the story after the flight. I said there's nothing that pleases me more than to have been so wrong—when I was one of two that told you you'd never fly again. It meant a great deal to him, and I think it was the right thing to do. I don't think NASA—or the people who selected the astronauts, including Deke Slayton or anybody else—needs to be ashamed for considering human beings when they make flight selections. As long as you don't select the wrong person. And I don't think that you have to select—what in a systems engineering, computer-run, paper organization would be—the ultimate best choice, because when you deal with people you can't . . .

I know of only one selection where if Deke had his own way, or for that matter probably JSC had its own way, it would not have been made, and that was with Jack Schmitt [Apollo 17]. I don't think Jack would have flown if it hadn't been pushed very hard from up here.

What about the capabilities of the spacecraft to land on land?

One of the things that I got into fairly quickly was that the spacecraft [Apollo] as designed did not have a land impact capability, that had been ruled out, and it was said "we aren't going to land on land." On the other hand, if you looked at the trajectory from the launch pad for some early aborts, unless you restricted yourself to only high off-shore winds, there were 30 seconds or so of land impact capability.

340

You may have heard in each launch that we had the call to the astronauts, "feet wet," which means that from that point on—this call was, say 15-20-25 seconds into the launch—we finally had overflown the line and any abort would make us land in the water . . . there may have been some exceptions. On almost every lunar flight—had we a mission rule which said you only launch when the wind is such, that you'll only have water landings, we never would have flown to the Moon. Because those conditions seldom exist. So we looked at the conditions at the Cape, the probability of a land impact, and decided that there was a high probability that we would have to launch under conditions when there could be a land impact. That was ruled out by a wave of the arm of Joe Shea's board. We looked at it, we undertook a very expensive program of proving, and deciding what had to be done to the spacecraft to make land impact possible.

I remember I called back all of the spacecraft that we had in museums at the time for the land drop tests. We went out to the Cape and measured the hardness of the soil when impacted by a spacecraft. We built a miniature Cape rig in Houston with sand that simulated the Cape. We tested it. We had some horrible pictures of those things. And we finally redesigned the amount of material underneath the spacecraft—between the spacecraft and inner shell. We redesigned the structure that held the hypergolic tanks. And we put different kinds of struts and foot restraints on the couches. In fact, we changed Wally Schirra's [Apollo 7] foot restraints three days before the flight. We also started a new wind-measuring technique at the Cape, and a computer program which would tell us exactly what we had and what the wind conditions were. And through all that we were able to launch under conditions where it never would have been possible if we ruled out a land landing.

Did you start this land-landing capability immediately after you took over from Shea?

I don't know. I remember one argument that I had with George Mueller was on this subject, because it was costly. I came in for one of the management council meetings, and among other things reported on the program that I was undertaking. I sought to understand the land-landing problem and to help fix it. And George said, "Well, what if you find that you can't withstand a land landing?" Well, we did find out what we could do. We were able to beef it up slightly, and through that we were able to ultimately write mission rules that allowed us to be much more flexible in flying. If we hadn't done it, two things could have happened. One is that later on Chris Kraft and the flight crew people would have prevailed on us and said, "Look, we aren't going to launch, or we aren't going to let you launch unless you have a water impact." And if we looked at that, as I said, with all the other launch window constraints on the lunar launching, we might not have launched toward the Moon. Or, we would have all gotten together and said, "Look, we will take the calculated risk that we kill a guy if we do have an early abort and land on land."

Looking back on Apollo today, that would not have been a bad risk because we didn't have a single abort. You can say, look we didn't have to make that change because we never had an early abort and we never landed on land, and therefore it was an unnecessary change. So you can look at any side of this you want to. But in retrospect, I think what we would have done is to compromise a little bit each way, and we would have written some mission rules which would have severely restricted our launch window and we might not have landed on the Moon before the end of the decade.

What are the things that really made Apollo go under your leadership—that made it possible to do the program?

The Change Control Board (CCB) and how it worked and the fact that I turned the whole software program over to Chris Kraft.

Chris had the mission control center, which had its own software but I'm talking about the programming of the onboard computers—the work that MIT [Massachusetts Institute of Technology] did. Chris Kraft and I, one time together, went to MIT to look at this system. I confided to Chris that I could not take on the added responsibility of making these work. It was just too big a job for me to also learn and understand the onboard computers as I'm working the rest of Apollo. I asked Chris if I could turn the whole job over to him. And I did, and he accepted it. The only thing I did with software thereafter is, once a week in my management center, meet with one of Chris' guys to review the schedules, make sure that the schedules were compatible with what we were doing. But Chris ran the Change Control Board. He worried all the changes, all the technical problems, all the astronaut interfaces, which were important and ran the show. It was just one problem. I gave away a third of Apollo to a guy that I knew could do it better than I could. And then I didn't worry about it anymore. And I'm frankly convinced that if I had kept it in my shop and had tried to do it, plus all the other things I was trying to learn and do, I couldn't have done it. So I started out defending CCB, but I think those two things together, the CCB and all the help I got from the Fagets and the Krafts and the Slaytons, and with Chris taking over the whole software package, were the two things that made Apollo possible.

ENDNOTES

1. After newly elected President Kennedy's victory over Richard Nixon in November 1960, Kennedy formed a "transition team" to assess the national space effort. That team was headed by Jerome B. Wiesner, a Massachusetts Institute of Technology physicist who would become Kennedy's science advisor. The "Wiesner Report" was released on January 10, 1961, and was highly critical of the quality and technical competence of NASA management, and of the heavy emphasis that had been placed on human space-flight. It called the Mercury program "marginal" because of the limited power of its Atlas booster and criticized the priority given to the Mercury program for strengthening "the popular belief that man in space is the most important aim of our nonmilitary space effort." "Report to the President-Elect of the Ad Hoc Committee on Space," January 10, 1961, NASA Historical Reference Collection. Quotes are from p. 9 and p. 16.

2. In the spring of 1959, a Research Steering Committee on Manned Space Flight was formed and chaired by Harry J. Goett of NASA's Ames Research Center. The Goett committee considered NASA's goals, beyond the Mercury program and by the middle of 1959, concluded that the appropriate objective for NASA's post-Mercury human spaceflight program was to send humans to the Moon.

3. Mercury-Redstone 2 (MR2) was launched from Pad 5 at Cape Canaveral, Florida on January 31, 1961. The spacecraft carried a 37-lb chimpanzee named Ham, a name derived from Holloman Aerospace Medical Center at Holloman Air Force Base, New Mexico which is where the Mercury animal program was based. The purpose of the flight was to qualify the spacecraft for launch aborts by firing off the escape rocket which carried the spacecraft away from the Redstone. The 4,190-lb. Mercury spacecraft no. 5 was lobbed to a height of 157 miles at a maximum speed of 5,857 mph and a downrange distance of 422 miles in a flight that lasted 16 minutes and 39 seconds. Ham experienced a maximum acceleration of 14.7 g due to a large overthrust with the Redstone, which threw the spacecraft to a greater distance than planned. Because of this, a booster development flight, dubbed MR-BD, was flown on March 24, 1961, with spacecraft no. 3, reaching a height of 113.5 miles, a speed of 5,123 mph and a downrange distance of 307 miles, fully qualifying the Redstone for the first manned Mercury flight.

4. In an attempt to launch the first Mercury-Redstone on a preliminary suborbital test of the spacecraft and launch vehicle combination before a primate flight, NASA attempted to launch MR-1 on November 21, 1960. However, the Redstone motor prematurely cutoff when the launch vehicle was about one inch off the pad. It settled back to its pad supports, but since the spacecraft received a cutoff signal it triggered its launch escape system and recovery parachutes. The launch escape tower broke free leaving behind the spacecraft on the launch vehicle with parachutes draping over the top shortly after the recovery equipment deployed itself. A second launch attempt, dubbed MR-1A, successfully launched on December 19, 1960. The spacecraft reached an altitude of 131 miles, a speed of 4,909 mph and a downrange distance of 235 miles before being safely recovered from the Atlantic Ocean.

Maxime A. Faget. (NASA Photo.)

CHAPTER 14

MAXIME A. FAGET

(1921 – 2004)

During his nearly 40-year career in government service, Dr. Maxime A. Faget has made many distinguished contributions to the advancement of aeronautics and astronautics. Internationally known as the chief designer of the Mercury spacecraft, he has played a major role in developing the basic ideas and original design concepts that have been incorporated into every manned spacecraft that the United States has since flown. From his early research in supersonic flight, through the design and development of the Space Shuttle, Dr. Faget's engineering acumen is evident throughout in the development of aircraft, missile, and spacecraft research and design techniques.

Born on August 26, 1921, in Stann Creek, British Honduras, Faget graduated from Louisiana State University with a Bachelor of Science degree in mechanical engineering in 1943.[1] Upon graduation, he joined the Navy as a naval reserve officer assigned to submarine service. Completing his military service, Faget became employed with the National Advisory Committee for Aeronautics (NACA) Langley Aeronautical Laboratory in 1946. He was assigned to the newly created Pilotless Aircraft Research Division (PARD), a division that flew rocket-powered models of aircraft and missiles at transonic and higher velocities to obtain aerodynamic data. It was during this period that Faget was first exposed to the idea of space flight while pioneering work on supersonic inlets and ramjets.

While at Langley, the NACA established a group to study and define problem areas that had to be solved to make space flight a practical reality. In March of 1958, Dr. Faget presented a paper called "Preliminary Studies of Manned Satellites—Wingless Configuration, Non-Lifting" at a conference held at the Ames Aeronautical Laboratory in Moffett Field, California. This significant paper put forward most of the key items that were later used in conducting the Mercury Project. It showed that a simple, nonlifting satellite vehicle of proper design could follow a ballistic path in reentering the atmosphere without experiencing heating rates or accelerations that would be dangerous to man. It also showed that a retrorocket of modest performance was adequate to bring this capsule down from its orbital speed and altitude to a reentry into the atmosphere. In addition, it outlined the possibilities of using parachutes for final descent, and small attitude jets to control the capsule in orbit, during retrofire and reentry. His paper concluded with a statement that "as far as reentry and recovery are concerned, the state of the art is sufficiently advanced so that it is possible to proceed confidently with a manned satellite project based upon the ballistic reentry type of vehicle."

When NACA became NASA in 1958, among the new organization's responsibilities was that of manned space flight. Dr. Faget was one of the original 35 members selected to form the Space Task Group (STG) which later developed into the Manned Spacecraft Center (now the Johnson Space Center). Although Mercury was the main task of the STG, there was great interest in developing follow-on programs. As chief of MSC's Flight Systems Division, Faget devoted a large amount of time to heading a design and analysis team that explored manned flight to the vicinity and the surface of the Moon. Because of this and other NASA studies, President Kennedy was able to commit the U.S. to a lunar landing by the end of the decade. With the advent of Apollo, Faget was appointed chief engineer at MSC where he was responsible for the design, development, and proof-of-performance of manned spacecraft and their systems. This responsibility also included specifying the function and design of numerous engineering laboratories to be constructed as part of MSC. In April 1969, shortly before the first lunar landing, he

348

organized a special preliminary design team to do an intensive feasibility study of a reusable manned spacecraft. This effort achieved program status when MSC was given the formal authority to develop the Space Shuttle. As a result, Faget focused the ensuing years on solving the numerous problems and technical challenges associated with the space shuttle until his retirement from NASA in 1981.

In 1982, Faget joined with several Houston businessmen and founded Space Industries, Inc. (SII). His company manufactured a wide range of experiment support equipment that has flown on numerous shuttle missions. The most significant of these was the Wakeshield Facility built for the University of Houston. This free-flyer was successfully deployed on two missions, providing experimenters with an ultra-high vacuum environment for materials processing.

Faget's numerous accomplishments include patents on the "Aerial Capsule Emergency Separation Device" (escape tower), the "Survival Couch," the "Mercury Capsule," and a "Mach Number Indicator."

Dr. Faget served as a visiting professor and taught graduate level courses at Louisiana State University, Rice University, and the University of Houston. He has also received numerous honors including the Arthur S. Flemming Award, the NASA Medal for Outstanding Leadership, and an honorary doctorate of engineering degree from the University of Pittsburgh.

Faget is married to the former Nancy Carastro and they have four children; Ann Lee, Carol Lee, Guy, and Nanette. The Faget's live in Dickinson, Texas.

Editor's Note: The following are edited excerpts from two separate interviews conducted with Maxime A. Faget. Interview #1 was conducted June 18-19, 1997, by Jim Slade, as part of the Johnson Space Center Oral History Project. Interview #2 was conducted on December 15, 1969, by Ivan Ertel and Jim Grimwood.

Interview #1

I believe in 1962 you were named director of engineering and development at the Manned Spacecraft Center, Houston. How did your day-to-day role change, or did it?

Well, it didn't really change. It was pretty much the same thing. The thing that was most important had to do with the move as opposed to the change in title. We came to Houston. We had to build a center. All of the engineering facilities had to be specified, worked out, negotiated, and an organization had to be built. We went from essentially a one-man, one-program project, [Mercury] to really trying to do three programs at once [Mercury, Gemini, and Apollo], plus build the center.

You were trying to telescope three levels of thinking—Mercury, Gemini, and Apollo. You were trying to do all this at once. Was that at different levels?

As far as Mercury was concerned, that was out of Engineering. We didn't have anything to do with it other than saying, "Yes, it's all right to fire, to leave the retrorockets on," which was about as much engineering as I did then.

And Gemini really didn't require an awful lot. There were a few new things on Gemini. There were the fuel cells along with their cryogenic oxygen and hydrogen storage systems. Also, Gemini was designed to be able to make significant translation maneuvers that required the use of a much more powerful auxiliary propulsion system. This was all carried in the adapter section that stayed attached to the capsule until reentry. Gemini also had an offset center of mass to provide a small lift vector so that it was able to maneuver during entry to minimize splashdown dispersion. You

Maxime A. Faget, assistant director for engineering and development, Manned Spacecraft Center, explains the function of the lunar excursion module to Father Patrick J. O'Brien, Mrs. Faget, and Mrs. Donald J. O'Brien, sister-in-law to Father O'Brien. This photo was taken on September 15, 1965. (NASA Photo 64-31226.)

know that Gemini was primarily the result of Bob Gilruth insisting with NASA Headquarters that it was essential to have more experience in space operations before we tried flying to the Moon. I'm convinced we would have never been able to make the landing in Kennedy's decade without the training and operations development that Gemini provided . . .

Apollo was the big driver as far as engineering was concerned. We had to get that under way with Apollo. No doubt about it. We just selected the program for Apollo and had to go with it.

And you were using Gemini as a testbed for so many of the particulars of Apollo, weren't you, the rendezvous and docking, the maneuverability of the spacecraft?

Rendezvous, docking, extravehicular activity . . . But that came about after it got to flying. The focus of modularization was still primarily on Apollo. We were going from a single vehicle to a two-vehicle system and had to come up with the specifications for the LM [lunar module], and do the whole procurement bit. That was a big effort. We'd actually, of course, done the procurement for the command service module while we were still in the [beginning of] Gemini . . .

We actually started looking at lunar flight, I guess, six or eight months after the birth of NASA. I remember Headquarters was trying to look at the future and understand what was going on. Well, one of the first things Dr. T. Keith Glennan did was create a group under Bob Gilruth to do the manned program. He put Bob Gilruth in charge of manned space flight. We didn't call it Mercury at that time. We still called it "the capsule." [Laughter]

So Gilruth had a couple of meetings, and then he went in and reported to Glennan. So NASA must have been about two or three weeks old. He told him where we were, that we'd selected this capsule shape and we were going to do this. Finally, Glennan said, "Well, that's fine. That's fine," he says. "Now what do you want me to do, Bob?"

Bob said, "Well, I think I want you to tell me to go ahead."

So he said, "All right. Go ahead. Do it."

So Bob Gilruth goes back, and he talks to the head of Langley, Tommy [Floyd L.] Thompson, and he said, "You know, that guy told me to go ahead, but he didn't tell me how. I haven't got any organization. I don't know how I'm supposed to do this."

They got their heads together, and Thompson said, "Well, why don't we just create the Space Task Group." So Tommy Thompson created the Space Task Group within Langley to do this job and put Bob Gilruth in charge of it . . . And then they got together, and they named 35 people . . .by letter that Tommy Thompson signed, that created the whole thing.

You say that you were basically most interested in Apollo when you came over to Houston. How much input did you have in the Gemini spacecraft? Was it just an extrapolation of Mercury?

It was an extrapolation of Mercury. It was not competed. What happened was, a man named Jim [James A.] Chamberlin came in. He's

352

the guy from Canada, from AVRO. He brought a bunch of Canadian engineers with him, which was really a godsend, because that put some real experience into the group. Bob Gilruth put him in charge of the day-to-day management of the Mercury Program. He created an organization and put me in charge of what amounted to engineering, although it was a different title, but it was essentially doing the same thing I ended up doing.

He and the people at McDonnell . . .were looking at all of the shortcomings of Mercury, because, as you pointed out, Mercury was not a vehicle that was controlled by people, but by the occupants. Mercury was a vehicle that just went up and came back down. It didn't do anything except stay in orbit until it was time to come down. They recognized that they wanted to do more than that, so they conceived of this program.

When Bob Gilruth told me about it, I said, "Well, they ought to have at least two people in there." That was my contribution: "They ought to have two people in there," so they made it bigger and put two people in there.

It's kind of interesting, Jim Chamberlin was really kind of the force behind the Gemini Program. There's just no doubt about it. Gilruth liked the idea of two people. He told McDonnell that they ought to have two people in there. Jim was the last one to find out we'd decided to put two people in there. [Laughter] But he thought it was a great idea, too. So we had enough to do. Titan was going to be big enough to carry two people.

That spacecraft was built by the same manufacturer as Mercury, yet Gemini did not use your escape tower. Why was that? Because of the Titan?

No.

Apollo did use your escape tower?

Yes. Well, that was somewhat of an aberration, and I argued long and strong against what they did, but they did it anyway. In the original concept, Gemini was going to make a land landing using a gliding parachute and they wanted to put ejection seats in it. In the event that something went wrong with the gliding parachute, instead of having a back-up parachute, they'd just eject. So they had the thing designed with ejection seats from the beginning. They said, "Well, we've got ejection seats. We don't need to put the escape tower on there." And there was a little bit of rationale there. The fact that they were using hypergolic propellant meant that the fireball would not be near as big. Now, you might say, "Gee, you're going to use hypergolic propellant where you just touch it and it's going to go off and you are going to have a smaller fireball?" Yes, that's the case.

The thing about liquid oxygen and kerosene is that they can mix quite a bit before they go off. You can't mix hypergolic propellants. The minute they start to mix, they go off, and for that reason they'd blow each other apart, and the amount that ends up getting involved in the fireball is very small. You can see if the tank were to spring a big leak or an opening, the propellants could mix quite a bit and then go off. Of course, then you'd have a really big explosion. So that was the thought.

But the bad part about the ejection seats, they probably would not have worked much over about 20,000 feet of altitude just simply because the velocity would have been too high. If you had to eject very quickly while the rocket was still firing in the back, for some reason or another, you were liable to go right through the fire from the rockets. The best thing about Gemini was that they never had to make an escape . . . Chris [Kraft] will tell you the same thing. If you ask him what he thinks about the ejection seats, he'll say, "I'm glad we never used them." [Laughter]

It seems to me that ejection on the pad with an impending explosion would probably have killed the astronauts. Or am I figuring that wrong, because it would have been a lateral ejection?

The ejection on the pad would have been quite marginal, no doubt about that. The parachute would open, and the man would make maybe two swings on the parachute before his feet were on the ground. You know, thank God we never did that . . .

Interview #2

I know you had a couple of meetings in September, and then on October 27, 1960, which defined the shape of Apollo. I'd like to get some of your thoughts leading to that date where you do get the shell of the vehicle outlined and some of the interior arrangement aligned.

Let's talk about the development of the shape. That's kind of a separate thing because as long as I get down that train of thought I'll think of some of the other things that went on at the same time. I don't know when the idea first came that we would go to the Moon and back. I do remember that one of the first things was the old military high ground approach—the military were the first ones who wanted to go to the Moon. They had funded studies on that. Once Mercury got started and we set aside two or three people— certainly by the end of 1959 we had two or three people—thinking about what might come next. Kurt Strauss was the key man working at that time on the Apollo program. Of course, Bob Piland was my assistant and a good bit of his time went on that.

When we got to thinking about going to the Moon we were faced immediately with the fact that reentering at 40 percent greater speed . . . at that time a lot of the entry experts started talking about hot gas cap radiation heat which is a very fancy way of saying that in the shock wave that stands in front of the entry body the molecules are first broken down into atoms and then further broken down so the electrons split off the atoms due to the energy of the air running in the shock wave. Subsequent to that, the atoms recombine, the molecules recombine, and the heat recombination is released in radiant heat. It is theorized that this radiant heat would be a predominating factor. As a matter of fact, early in the Mercury program it was considered that maybe the radiation heating might be a predominating factor even at the entry velocity of Mercury. We realized very shortly in the program that after Big Joe,[2] and a few flights like that, this wasn't going to be the case. But that just allowed the theoreticians to encourage them to move their velocity of concern up a little higher.

Now, it turned out that the blunt body is best to protect against the conductive heat transfer which is the nominal heat transfer of heat from the vinyl area of the surface, whereas the blunt body is the worst offender from the standpoint of this radiation heat transfer because the blunt shock wave that is formed—the blunt body—maximizes the strength of the shock wave. The stronger the shock wave is the greater tearing up of the molecules and, consequently, the greater would be the subsequent recombination in the radiation. The whole business of estimating this type of heating is based on analysis that required assumptions that you would not be able to verify by experiments. Estimations of this radiation heating varied over a factor of about 20—from the worst case, to practically none at all. Of course, if we'd taken the worst case, the blunt shape—such as we used in Mercury—was not a very good shape to use. The result of this was to cause a reconsideration of the basic shape.

Two main contenders were the M-1 shape, which was the shape that Dr. [Alfred J.] Eggers, Jr. had designed which was a half cone with a very small blunt base to minimize radiation heating. Another one was the lenticular which would come in edgewise because that had a low heating, and GE had a shape which was kind of a conical shape. There was great concern over the amount of lift over drag (L/D) that would be required during entry. The basis of this concern was due to uncertainty of the entry navigation. Dr. Harry J. Goett was probably one of the most pessimistic of the group. Frankly, he was completely uncertain that we could navigate back from the Moon. Depending again on the estimation of how accurate one could navigate back into the earth's atmosphere, i. e., how close to the center of the entry corridor you could hit, was a great determining factor on how much drag you needed. If you could hit close to the center of the corridor, lift to drag—high lift to drag capability was of benefit because it would allow corrections in navigation areas before either the heating on the G's got too high on the one hand, or on the other hand before you skipped out. You could pull negative lift and hang in there. The Mercury type shape, blunt heat shield shape, was admittedly limited to about L/D of a half, and a lot of people felt we at least needed L/D to one and some of them were suggesting L/D of one and a half. That's about as high L/D we could have. The higher L/D, the lower the drag and consequently the lower the efficiency as a reentry body. So we had a lot of things going.

The people at Ames were the first ones to come up with some fairly good analyses that showed that the hot gas gap radiation wasn't as severe as the worst predictions. [Clarence A.] Syvertson was one of the guys who did a lot of work to that end. Then we had started doing some studies with MIT [Massachusetts Institute of Technology] and got a lot more confidence in our reentry navigation. So by the time that we were in the middle of these three funded studies, we weren't overly concerned about the hot gas gap radiation. The worst that the blunt body type would experience would negate its basic advantages in low heating.

On the wing vehicles, the high L/D's were not needed anymore. I guess I got bold enough to tell them that. I had a feeling that was right all along. It's one thing to tell them, but it's another thing to tell them with authority that they'll believe you, and I think we gained their confidence that we didn't really punish ourselves heating wise by going with the very blunt vehicles. We stood a chance to come out ahead—at the worst situation on radiation would be about an even trade off comparing a blunt body to some of the more pointed bodies. On the other hand, we'd come out way ahead. It turned out that the lowest estimate we made was the one we actually experienced.

During the time of the studies, we set our own people to work and we had two concepts in-house, one was the M and M shape or the lenticular vehicle, and the other was the derivation of the Mercury which was like a Mercury shape only it had a shorter afterbody because we planned to fly it at a high angle of attack—we didn't want the afterbody to be exposed. We made the conical angle blunter back there. At the same time that we were beginning to get some favorable answers on the heat radiation and some definitive analysis on the accuracy of the return navigation, and we felt pretty certain . . . and this came not only from MIT but from our three study contractors. All of their navigation analyses indicated that we stood an excellent chance of being able to get in a very narrow corridor, so now instead of just having Harry Goett's conservative feelings that we might not be able to make it, we had some very definitive studies that indicated that state of art navigation supported a very good chance of getting into the 10-mile corridor. Of course, if anybody said now you had a 10-mile uncertainty, it would sound terrible the way we're navigating these days. Mind you, we were trying to prove that things weren't like 100 miles. Ten miles was good enough to support the L/D of a half vehicle. So along about the middle of the study period we put the word out that we thought the L/D of a half was adequate for entry navigation and there was no necessity to be overly concerned about the hot nose cap radiation.

Meanwhile, our own in-house studies ... looking at both lenticular and the Mercury derivative shape, the main reason for the lenticular shape was to try to get a L/D of an excess of one hypersonically and the interest in that shape decreased when we found we didn't need that high L/D. It also looked like it was not as competitive from a standpoint of heat protection system as a low L/D.

Radiation also influenced the progress of the design. I guess there was a number of configuration drivers in addition to just the reentry shape that we were concerned about. We decided early on that we were going to make the thing big enough. We were just at that moment experiencing all the bad things that came out of making Mercury too small. One of the things we set as a policy in all our design studies was an adequate amount of volume inside the command module. One of the other things that was being studied was the possibility of a two-compartment vehicle as opposed to a one-compartment vehicle. Now I'm not talking about LM's or anything like that. The mission during this period was merely to go into orbit around the Moon or just circumnavigate the Moon and back. We were looking at about a one to two-week long mission and it would terminate with the reentry. Now in order to have enough volume, of course, they had to make the thing bigger which meant we had to carry along a lot of extra heat protection systems, so it seemed a very attractive thing to divide that volume in two pieces. We had for a long while a command module and a mission module, the mission module being where everybody was supposed to do their business.

This started off to be a very attractive idea but as we went through our own studies and the contractors went through their studies it became clear that less and less things were going on in that mission module, and everything that was vital for one reason or another also was vital during entry so you either did it twice, once in the mission module and again in the command module, or you did it once in the command module. So it seems that the mission module was turning out to provide nothing but extra room. There were no systems and no particular activity that anyone really wanted to carry out in the mission module other than to stretch out and perhaps get a little sleep. The consequence of this was that it didn't look like it was worthwhile to have a mission module. So in the final analysis we ended up with a single cabin version. You might have noticed that the Russians ended up going into something very close to our two-compartment vehicle that we were considering. I don't know where they got their ideas, but it might have been from us. We made no secret of these considerations.

The entire layout of the vehicle became very important at that time and we had a couple of people make very detailed designs. I think the most notable people here were Owen Maynard, Larry Williams, and Will Taub. There may have been others but I don't recall them. Of course, Caldwell Johnson was leader of that group. Will Taub, I think, constructed a 10-scale cardboard mockup of the vehicle and we also constructed several full-scale plywood mock-ups of the vehicle during that time. The purpose of all this was to make sure there was enough room for everybody inside and that everything would work together. The layouts that we ended up with are very definitely derivatives of that early work. The command module in many ways is pretty much laid out the way those fellows said it should be . . .

There were some other things that we did at that time that helped set the shape. We had a fairly lengthy argument about whether the bottom of the heat shield should be rounded—the point where the heat shield and the conical part intersect. Mercury and Gemini are sharp and in Apollo we rounded that off. I thought it should be sharp because I couldn't see anything wrong with it and it also increased the total drag.

One of the primary things that I think settled the issue was that on the early Saturns, which had been designed primarily as a R&D vehicle down at Huntsville, they designed this vehicle and got the go ahead to put two or three of them into production for test flights without any payload. So they had several old Jupiter nosecones laying around. These are long conical nosecones, and they decided since they had instrumentation in those nosecones they'd make the Jupiter nosecones the payload. But because there wasn't enough room for all the instrumentation in a Jupiter nosecone, they discontinued the conical shape until the diameter got up to 156 inches and said that's going to be the payload. Then they took the upper part of the Saturn IV-B stage, which was being built by Douglas, and put an adapter section on it at the same conical angle as the nosecone and brought that down from the 180 inches to 156 inches. That was the interface between the work going on at Douglas and the work going on at Huntsville. So when we finished our job we had a vehicle that was 14 feet in diameter and we said we'd like to fly this on top of the Saturn and they said "Oh, gee, it'll cost us a million dollars to put a new front end on the Saturn." So we went round and round, and to make a long story short we decided at Langley that it would be a lot easier just to reduce our diameter down to 156 inches than to argue with those guys. And the easiest way to do that was just to round the corners off—we didn't want to change the internal layout, so we rounded off the corners and ended up with 156 inches. Marshall decided that they would put rounded corners on Apollo and I don't think they ever knew why.

Another thing that was settled concurrently was the type of heat protection we were going to use. We certainly were going to use an ablative heat shield in front—on the blunt end coming in. There was a lot of argument on whether we should use an ablative material or shingles on the conical portion. We had used shingles on Mercury and Gemini.[3] I thought we ought to use shingles on Apollo, and a lot of people thought the heating would be much worse on Apollo; it was coming in faster and we wouldn't be able to handle it with the shingles. They felt we should put ablative material back there. Well, the ablative material was definitely going to be heavier than the shingles and we were concerned about making that decision in as much as we didn't have any hard data to go on. But the thing that swayed me to the ablative material was that all of the analysis we had done up to that time showed that the occupants inside would get a pretty severe radiation dose during the mission. The additional weight would have to be included in the command module just to provide protection against solar radiation. So I thought there was very little sense to try to make the skin thinner and then have to put in a lot of weight inside to protect against this radiation. On that basis I stopped arguing with the people that wanted to put an ablative heat shield on the conical portion. I guess that was a mistake—I should have stuck by my intuition a little longer because the radiation problem wasn't near as severe as the people like Dr. Van Allen thought. He was very strong in his views that the radiation problems would be very severe. Neither situation turned out the way we designed for. The heating on the back has been very, very mild—we could have gotten by with the same kind of shingles used on Mercury and Gemini on the back. We didn't need all that material back there as protection against radiation, so perhaps we had some extra weight in the vehicle as a result.

How did they go about determining reliability factors for Apollo?

Nick [Nicholas E.] Golovin worked for NASA then and he had a background in mathematics. He became Mr. Reliability for NASA and proceeded to take a strong mathematical approach to reliability. His basis was that by use of proper statistical analysis you ought to be able to predict what the reliability of a system is. You do this by creating a network of all the components in the system that can fail in the proper mathematical model, and then you estimated based on experience during development, testing, etc., what individual reliability these components in this big mathematical model vehicle is. Then you just apply the proper amount of mathematical machinery to it and out comes the overall reliability. He fell very much in love with that system and there was no doubt about it, he thought it was great. I think where the engineers got a little bit uncomfortable with it was they really didn't know how you got the basic data. We thought we ought to use all of the test work up to that time. We felt the reliability performance of a system during its development phases was not much of an indication of the ultimate reliability of any component. We went round and round on this. Nevertheless, a mathematical reliability number was created for Mercury and I don't know what went with it. One of the things that happened in Apollo was everybody felt like we've done this in Mercury, and reliability never really influenced the design of Mercury, which, of course, really isn't so. It was designed to be as reliable as we could design it, but the thought was we ought to start from scratch in Apollo with some reliability goals so that the engineers would be properly motivated to reach adequate goals for reliability.

I remember early in the program I asked Walt Williams what he thought reliability ought to be for Apollo and he said, "Well, there ought to be at least three nines on safety and two nines on mission success." So we said we'd try that number. And that's the number that went in. Shortly after we'd put that number out one of the study contractors came back to me and pointed out that this wasn't very different from the expected mortality from three 40-year-old individuals on a two-week mission if you took the standard actuary tables. But anyway, that's the number we used and I guess no one can argue with the achievement.

There were two things that we made exceptions of and I think we did well on both of those. This came later in the program during the North American early development phases. One was the reliability or I guess you might say the invulnerability number to meteors, and the number with respect to radiation dose. Any analysis made during the early phase of the program—these two hazards predominated the situation. As a matter of fact, the three nines in reliability left nothing over for system failures, because all the failures were going to be from meteors or from a guy getting an overdose of radiation. What we did was essentially said the reliability model wouldn't consider these two hazards, that we would try to achieve the three nines in reliability without it and there would be added considerations. We set separate goals for meteor penetration and radiation dose and I don't remember what the goals were. We did set separate goals for those. I might say that since that time

Members of the Apollo 204 Review Board, Astronaut Frank Borman and Max Faget (foreground), inspect the interior of an Apollo Command Module mock-up at the Kennedy Space Center on April 9, 1967, as part of the Board's investigation of the fire that took the lives of the Apollo 1 crew, Chaffee, White, and Grissom at Launch Complex 34 on January 27. The mock-up interior was established as closely as possible to the configuration of the 204 spacecraft before the fire. (NASA Photo 67-H-376.)

up until now both of the hazards from these two environments appear to be a hell of a lot less. We did a rather sensible thing.

On Gemini X they were still saying these hazards were going to be high.

That's right, but we had an additional problem—we were dealing with radiation outside of the Van Allen Belt and it was pretty much unexplored from the standpoint of total radiation and we had the solar storms situation where you get these great big solar events. It's very difficult to predict how severe they would be. So the whole thing was completely unmanageable from the standpoint of design engineering. We did, for a long while, carry the idea that we would provide about two-inch thick Lucite goggles for the crew because the most sensitive portion of the crew to radiation was the eyes— so in the event we got into a solar event the crew could cover their eyes with these thick Lucite goggles, Lucite having nice low density atoms that stop the radiation. But we don't carry those now, apparently the radiation hazard has decreased some more.

Another thing of big concern early in the program was whether or not to use pressure suits and what to do about personal parachutes, escape seat ejection. In both of these cases my group was recommending that we not provide these features. We made a recommendation not to use pressure suits inside the vehicle. We didn't think they were necessary and we felt that the same weight that goes into the pressure suit and the system in the vehicle to support the suit, the extra hoses, etc., could better be used in making the cabin safer. The operations people, primarily Flight Crew people and Walt Williams didn't really think that was the safe way to fly. So we designed a vehicle to be flown with the suits on all the time. Since that time we have learned through experience that suits aren't really necessary. It took the crew to decide against the suits as opposed to program management.

Remember the public argument on Gemini VII?

You see, in Gemini VII it was the crew that wanted to take the suits off and in Apollo it was the crew that wanted the suits on. The big voices for suits on were the crew people and Walt Williams who, of course, represented the crew to management at that time. We had a similar thing on bailout parachutes. The crew thought they would like to have chutes inside the vehicle for getting out. We went through quite an exercise on this. After North American was awarded the contract for Apollo, their original design incorporated a much larger hatch—where the quick-open door is located now—that explosively released in case of an emergency. This larger hatch spanned three couches and each crewman would have his own chute. The idea was, in the event the main parachutes weren't opening or malfunctioned, or some other trouble happened, they could blow the hatch and get out. We went through quite a few exercises with mockups trying to decide how long it would take to get out and finally decided that the cases where this would really be effective from the standpoint of saving their lives was so small that it wasn't worth carrying the system. But we did initially start off with the idea.

In addition to that, I might mention that we had a lot of design efforts that went into the landing impact system. I think the statement that was once made about how Apollo carried its landing gear inside is pretty much the case. We ended up deciding on putting the couches on shock struts and stroking the couches on impact as opposed to putting a landing bag like we had on Mercury or, in the case of Gemini—at least during the early days—we had real landing gear. I'm not so sure that was the right thing in retrospect either, but it got in there initially. It was reconsidered a number of times as different people from time to time made all sorts of arguments for some other systems. The expediency of the situation that here we had something in hand always prevailed as opposed to the cost and the schedule impact of slowing down and going with some other approach. Undoubtedly, the better thing to have done from the very start would have been to include small landing rockets. If we had it to do over again, we certainly wouldn't put the landing gear inside—we'd put landing rockets on there instead. We studied rotors, we studied gliding parachutes, we studied rockets up in the parachute riders, landing bags, all sorts of things as alternatives. Of course, we studied three parachutes, six parachutes and sequential parachutes.

I guess one of the biggest decisions made during the Apollo program was when we decided to land on the Moon. We started off with the idea that we'd build a vehicle that would be able to circumnavigate the Moon, but we always put in fine print that it would have the development potential of landing on the Moon. I think from the very start we always thought of the command module itself landing on the Moon as opposed to using the LM. It turned out, when it went out on contract for the studies—the three studies by Martin, Convair, and GE—they were all for circumnavigation. When they finally got the go ahead from the President to make a lunar landing, we had brought the program far enough along that we just amended some work statements and went out for that idea and everyone bid on an Apollo type command module actually landing on the Moon. This called for a very, very large rocket to include a landing. One of the interesting things about that thing was that the take-off rocket that was used at that time was about the same size as the service module—the same total impulse. So although at the time we actually mailed out the request for proposals we had already done enough looking into this lunar orbit rendezvous mode to realize that we could save the service module—at that time, of course, it was the return rocket, that the return rocket was the proper size for the service module. Nevertheless, when North American got under contract they thought they were contracting to build the vehicle that was actually going to land on the Moon, which proved to be a little troublesome some time later . . .

Everybody had a plan. Very early in the program we had the Goett Committee, the Low Committee, and groups in between. Near the end of the Goett Committee there was a meeting at Headquarters, prior to the time of the formation of the Low Committee that Dr. Glennan called. The thought was what to do if we want to do more than just circumnavigate the Moon or orbit the Moon. How would we go about landing on the Moon? Both Dr. Pickering and von Braun were there and this is when we were using the horizontal . . . the thing that landed on its side has basically got a lot of stability to offer as opposed to the thing trying to stand up on landing. In order to the make the landing, I showed a chart at this meeting which showed that a landing could be made in two phases; one, we'd go into orbit around the Moon and after getting into orbit we'd, two, come down from orbit to the lunar surface. Dr. von Braun questioned this right away. He said, "Well, why do you want to do it that way? We're already developing unmanned vehicles—the Surveyor in making a soft landing on the Moon, it doesn't go into orbit, it comes right straight in." Dr. Pickering got up and made a little speech about they had all the techniques worked out to that approach and you don't have to go into orbit if you want to land on the Moon. You just aim at the Moon and when you get close enough turn on the landing rockets and come straight in. I mentioned to them "that would be a pretty unhappy day if when you lit up the rockets, they didn't light." I got the feeling that this one session wasn't enough to convince them . . .

I only mention this to indicate how primitive things were at the time. There were some very well respected individuals in NASA who seriously thought we ought to go right to the Moon, and when you get about so far away from the surface, turn on the brakes and land. The advantage of going into lunar orbit . . . this kind of basic thing, had not even been considered. As a matter of fact, the outside advice was the other. The advantage of going into lunar orbit is, of course, you're not committed to anything—when you go by the Moon you have a free return, and when you do go into lunar orbit you're not committed to landing. You're only committed to landing when you've lit the final burn. The mechanics of doing it that way are basically safe and it minimizes the commitments until you're certain of the capability to do the job whereas the other approach did not. That hadn't occurred to these people prior to that time.

Like I say, the thinking at that time was very primitive and certainly explains why we had the big hullabaloo we did over LOR, EOR or direct. When we started studying the LM it was presented primarily as a method of saving weight and a method of reducing the total amount of weight. I was pretty much antagonistic toward it at first, primarily on the basis the weight estimates were in error. At that time, we needed about 120,000 to 150,000 pounds to go in direct, and to go with the LM there was less than half that much.

The earliest versions of the LM were . . . I think they weighed 2,000 lbs., and it was in that period of time that I was very concerned about whether we'd do that. Part of the problem there was we had two different groups making weight analyses. Well, I guess I did raise some objections to the LM based on the fact that I thought the people trying to sell it hadn't done a very good job on

During the Apollo program, one of the projects Max Faget directed was the lunar communications relay unit and ground command television assembly used for the first time on the Apollo 15 lunar landing mission. This equipment greatly improved communications between the Earth and the Moon and allowed people around the world to watch the lunar excursions real-time on their televisions. For the work done in this area, NASA received an Emmy Award from the National Academy of Television Arts and Sciences. Shown here is Faget accepting the award on behalf of NASA on June 29, 1970. (NASA Photo 70-44081.)

estimating what it weighed. When it came down to really doing the job there was no doubt about it, the best thing about the LM was that it allowed us to build a separate vehicle for landing—it didn't have anything to do with lighter or heavier . . .

The decision was made that the approach that MSC [Manned Spacecraft Center] initially wanted to take was to go "doll up" mode, so called, where we would build a Saturn that would lift

120,000-150,000 pounds. With that size Saturn, we felt we could land a command and service module itself right on the Moon, which would put all three people on the Moon . . . we thought that's the way to do it. It also had the advantage that if we were doing that now and wanted to stay a week we'd have no trouble staying that long. We started down that path and the first thing that happened was that we had the Golovin Committee get involved in designing the next program boosters. We got Huntsville . . . I forget how this all worked out but we ended up with a five engine Saturn instead of an eight engine Saturn. Wernher [von Braun] first backed off from eight engines to four engines on the basis that this was as big as he could build it. Then he ended up putting that middle engine in there because it seemed like it would fill up the hole . . . Nevertheless, the stated reason we couldn't build an eight engine Saturn was that we were going to build it at Michoud [Michoud Assembly Facility at New Orleans, Louisiana], and the biggest tank they could build at Michoud was 33 feet in diameter, and that went with five engines, not eight. If we wanted to go to eight engines, it had to be something like 36 or 38 feet in diameter. It just wouldn't fit without raising the roof at Michoud. I think they basically wanted to do earth orbit rendezvous at Marshall, and they felt the best way to get the smarts to do that was to make sure the booster didn't get so big they didn't have to. That's dirty thinking I guess, but I really think that motivated them a hell of a lot—they said "well why should we worry about making it so big, we can always do earth orbit rendezvous." What came out of all this high-powered planning was that we would build a five-engine Saturn, and we would have earth orbit rendezvous. That was the way NASA started down the line and, of course, that's the way North American started off on the contract.

In the case of actually doing earth orbit rendezvous two problems came up, one was do you do earth orbit rendezvous by fuel transfer or do you do earth orbit rendezvous by actually hooking things together in earth orbit. As far as I know, that problem never got solved. Every time you'd tell them what was wrong with one way of doing it they'd tell you well they were going to do it the other way. The business of rendezvousing in earth orbit was not the problem, but the business of after you rendezvous in earth orbit how do you put together the wherewithal to go to the Moon. That was the problem, and that problem was never solved.

There was another thing that bothered us, at least it bothered me, and that was we really started off designing a vehicle to do lunar orbit. In order to make that thing land on the Moon, the original plan had to make the landing stage use hydrogen-oxygen propulsion and RL-10 rocket engines. The RL-10 had already demonstrated that it could be throttled . . .

It turned out that the decision was to give the Lewis Laboratory the landing stage for Apollo, and . . . I didn't think a hell of a lot of it . . . it meant that instead of two Centers doing the Apollo program it would be three Centers. I always felt that the landing stage would be a pretty intimate part of the vehicle. The interface between the landing stage and the rest of the vehicle was a lot more intimate than the interface between spacecraft and launch vehicles. It was a much more intimate interface that we were going to have to make two Centers work on. So, after looking at the problems in this thing, the crew would have to . . . well, you had all the landing dynamics, the landing gear, the flight control right down to the lunar surface, and all that . . . it looked like it would be a big thing. We did some more studying and we were able to show that if you really wanted to land on the Moon it probably would be better to separate from the hydrogen-oxygen system just before you land, and land on a real short stage. This had to do with the height of the landing stage and a bunch of other things. So we generated what we called the lunar crasher concept, which meant there was a hydrogen-oxygen stage that was used first to get into lunar orbit and second to descend to a couple of thousand feet above the surface of the Moon, and then you go into the hover mode, or the throttle down mode. At that time, we jettison the hydrogen-oxygen stage and lit an earth storable propellant landing stage using the same kind of propellants we now use to land. The vehicle that actually lands is a lot more compact and would have all of the good features that you obtain by a pressure fed storable system. It also had the added benefit, at least to us, of greatly simplifying the interface because then we could consider the hydrogen-oxygen stage like just another launch stage. It didn't play as intimate a role into the maneuvers . . . it didn't have the landing gear and all those things that interact with the rest of the vehicle as deeply as that would. So we went to Headquarters and argued long and hard for the lunar crasher, and as a matter of fact we did sell the concept. Lewis was then under direction to build the lunar crasher stage, and we were going to have North American build the lunar lander portion. I really didn't like the idea of earth orbital rendezvous. Not liking earth orbit rendezvous was one of the things that made lunar orbit rendezvous look very attractive. It got rid of a hell of a lot of problems that you can see were being generated.

I've heard it said that you were the second man who was ready to go LOR. I think I'm beginning to see the reason.

I think that Bob Piland and Kemble Johnson were against it. Gilruth told us one day, "We really ought to study that real good." No doubt about it, he had a lot of influence on it, but we started perceiving these other problems, the problems of earth orbit rendezvous, and the problems of the lunar crasher, and the interface with Lewis. Boy, lunar orbit rendezvous really looked like the thing to do. In addition, it did get us out of a very difficult thing which we never solved, which was how in the world were we going to fly that command module down to the surface of the Moon. We had all sorts of little ideas about hanging porches on the command module, and periscopes, and TVs, and other things, but the business of eyeballing that thing down to the Moon didn't really have a satisfactory answer . . . No doubt about it, we'd have had to put a porch on that damn thing so the guy could sit up there and look, and once we'd provided him with that I think we could have gotten it down, but it would have had a lot more complications. And of course in the case of the command module, it would have been way up there—we'd have had to put a long ladder down. It would have taken quite a ladder. But, we would have ended up with three guys on the Moon.

Interview #1

There were two episodes in Apollo which were calamities, one of them not so much as the other. The first one, the Apollo fire on the pad, what happened there? What was the pressure inside NASA that created the need to move into the program with a Block I spacecraft that you knew was not going to be the flight model, the full flight model? What caused that particular circumstance to develop?

Well, we had Block I and Block II. Block I was in manufacturing long before Block II was, obviously because it was I instead of II, and it was getting close to completion, and it just made sense to go ahead and fly it instead of waiting for Block II, because you had to make progress. There were a lot of things wrong with Block I, but the main thing that was wrong with Block I was not something that was anticipated in Block II, namely that there was too much flammable material aboard [and] that we didn't properly recognize how fast a fire would propagate and that there was not a way to get out. In other words, this hatch couldn't be done from the inside. Once the pressure started building up, it was glued in there.

In a pure oxygen atmosphere.

Yes. Hindsight is wonderful. We had the same atmosphere in Mercury and Gemini as we had in Apollo. They never had any fires. But, you see, after I started thinking about it, kicking myself for being so stupid, I realized that the difference between Mercury and Apollo was that one Apollo experience was probably equivalent to maybe 20 or 30 Mercurys, simply because there's so much more volume in Apollo and there's so much more stuff in Apollo, so that it's going to burn just as badly. It only takes a teeny bit of stuff, with some teeny bit of flammable material to ruin the whole thing, but there's so much less material in Mercury and so much less in Gemini, actually, than Apollo, that the odds of it happening in Apollo—you'd say, well, sooner or later it would have happened in one of the Apollo flights. It just happened to happen on this one.

As you probably know, I was on the review board after the fire. We never did find out what caused it, what specifically caused it, but the real reason it happened is that we had too much flammable material in there and we had a completely pure oxygen atmosphere. Now, one of the things that resulted from that is, we worked on both. We had a very extensive program where we actually tested the flammability of everything that went in there, and we coated items with nonflammable stuff.

Did you know we came to an impasse, though? We found out that we could not completely be assured that we would not have a fire, in spite of all the changes we made to the material, and that's simply because we had an atmosphere which was about a half a pound of pressure greater than sea level of just pure oxygen in there. You say, well, why did we have the oxygen atmosphere in there? Well, it's a complicated story, but it's one that we're pretty much trapped in. It all gets down to the fact that we wanted to be able to go in and out of this vehicle in a spacesuit at any time without any pre-breathing, so we could not afford to have any nitrogen in the air if we were going to get the man down to possibly three psi, which is the lowest pressure that you might end up with when you go out in a suit. It doesn't take very much nitrogen in the atmosphere to give you the bends, if that's what happens. So we start off with the oxygen.

Well, I got to thinking about this, and I said, well, you know, you're not going to go out in a suit for a couple of days after we launch. So we went to what we call the 60-40 atmosphere, which would solve the problem. Just putting that much dilution into the atmosphere, 40 percent nitrogen, greatly reduced the flammability of a lot of things that we just couldn't make inflammable in tests . . .

I think Block II probably would have been just about as flammable as Block I. Now, there were other things that were not good in that spacecraft when you looked at it. The wiring was not too well done. It was not neatly done. There were too many fixes and so forth in the wiring. Where they found something wrong with the electrical service, they'd tape over or jump over it. A lot of circuits were added later on that weren't anticipated, which made for junky wire bundles. Obviously neatness is an important thing when you think about fire hazards. The spacecraft itself was not too neat in Block I as a concept, and all this was cleaned up in Block II.

Do you feel, in hindsight, that the agency was pushing too fast in getting ready to fly the Block I spacecraft?

No. I didn't think so. If they'd said, "Relax, take another three or four months," we'd still have probably flown the same spacecraft, still would probably have run the same tests, still probably had the same goddamned fire. If we had waited until Block II, it might have been a little cleaner, but I'm not sure. There was an awful lot of stuff in there. They were using Velcro all over the place and they were patching up. Papers were here and papers there. It was just a relaxed attitude towards fire, which was not called for.

Now, we gave Rockwell the title of "fire marshal," you know, which is kind of like the guy who comes in the theater and looks around and says, "No, you can't have people in here. You've got this, that, and the other you've got to do before I'll allow this auditorium to be occupied by a big crowd." They had not done that yet. I'm sure they were going to do it, and, probably, given enough time, they probably would have done it, but I don't think it was high on their priority list.

What was the biggest challenge to Apollo?

The biggest challenge to Apollo, I really think was propulsion. When astronauts are in Earth orbit, they can come down so easy because all they've got to do is slow down a little bit, and they're going to come back into the atmosphere, and once a vehicle has been through an entry aerodynamically and you know it's controllable, you don't worry about it burning up. But when you're on the Moon, you're in a gravity sink. Your propulsion system has got to work. It has really got to work. You can't wish your way out of that sink. [Laughter] You know, being in lunar orbit's one thing. You've got to come up with something like 2,000 or 3,000 feet a second to get out. On the surface, you've got to come up with a lot more than

This early illustration shows a concept for Apollo which would have employed either the direct ascent or Earth orbital rendezvous mode of operation. Shown here is the all-up configuration which allowed three astronauts to travel to the surface of the Moon for up to a week's stay and return using a single spacecraft. Note the towering height of the vehicle, its return upper stage and the now familiar command module which the crew would have used for their return to Earth once the upper return stage vehicle blasted off from the lunar surface. (Photo courtesy NASA.)

that to get out, and, of course, you've got to get up and make the precision of the rendezvous.

What you would call the mission analysis guys had to do a lot to make sure that they could abort anytime during the descent and still rendezvous. We carried a lot of propellant aboard for contingencies, and as we developed the flights and began to better understand, we gave up a lot of that contingency capability because we realized it wasn't necessary.

When we first started thinking about the lunar program, the question was how much propellant did you have to have to correct for navigation errors [on the way out and during the return], how much spare propellant? If you make an error, the only way you can correct the error is to change the velocity, and the bigger the error, of course, the bigger the change in velocity. We were carrying something like 500 feet a second of propellant, just reserve propellant for errors in the first analysis of the lunar mission. By the time we actually got to flying, we were down to maybe 100 feet a second. I think on some of the last flights we used maybe two or three feet a second on the way back from the Moon to correct errors. [Laughter]

What you find out is that you can really track good and you can detect error when you're at the Moon. Say you've got a tenth-of-a-foot-a-second error at the Moon. If you don't correct that, by the time you get maybe a half an hour from hitting the atmosphere on Earth, that tenth of a foot a second could be twenty feet a second in correction that you'd need. So as our tracking capability improved, we could detect errors early, correct them early, and correct them with assurance. It's not that we were just burning propellant because we think there's something wrong; we know what to do, we know how much to correct. You really don't have to carry all that contingency.

ENDNOTES

1. Several other members of what would become NASA's Space Task Group also attended Louisiana State University. Paul Purser earned a B.S. degree in aeronautical engineering from LSU in 1939 and worked briefly for the Martin Company in Baltimore, before joining Langley that same year. Joseph G. "Guy" Thibodaux and Max Faget were college roommates and both graduated in 1943. In looking back on those days, Thibodaux recalls "The interesting part of it is that Paul Purser, Max Faget, and I were all LSU graduates. Max and I were college roommates. We (Max and I) had a pact that at the end of the war, if we both survived, we'd get together and go look for a job together." Excerpt from an interview with Guy Thibodaux conducted on September 9, 1996, as part of "Space Stories: Oral Histories from the Pioneers of America's Space Program," an oral history project conducted in conjunction with the Houston Chapter of the

AIAA and Honeywell Corporation. The interviews were conducted by Robbie Davis-Floyd and Kenneth J. Cox.

2. On September 9, 1959, at 08:19 UT, the first major test of the Mercury program began when an early production Atlas D was launched from Cape Canaveral carrying a boilerplate spacecraft 95 miles into space and 1,496 miles down the Atlantic Missile Range in a 13 minute flight. Programmed to reach a speed of 16,800 mph and throw the Mercury spacecraft 2,000 miles downrange, the two Atlas booster motors failed to separate at 2 minutes into the flight and, encumbered by the additional weight, the vehicle achieved a speed of only 14,857 mph. The spacecraft experienced 12 g during reentry, and when it was recovered the next day, inspection of the shielding revealed that less than 30 percent of the ablative material had eroded, despite higher-than-planned-for temperatures during reentry.

3. The shingles used on Mercury and Gemini were made of Inconel-X, a nickel-chrome alloy. This material had been previously used with great success on the X-15 program as it was "capable of rapid heating to high temperature (1200 F) without developing high thermal stresses, or thermal bucking, and without appreciable loss of strength or stiffness." Loyd S. Swenson, Jr., James M. Grimwood, and Charles C. Alexander, *This New Ocean: A History of Project Mercury*, NASA SP-4201, GPO 1962, pp. 57, 63.

AN ANNOTATED PROJECT APOLLO BIBLIOGRAPHY

Armstrong, Neil A. et al. *First on the Moon: A Voyage with Neil Armstrong, Michael Collins and Edwin E. Aldrin, Jr.* Written with Gene Farmer and Dora Jane Hamblin. Epilogue by Arthur C. Clarke. Boston: Little, Brown, 1970. This is the "official" memoir of the Apollo 11 landing mission to the Moon in 1969. Contains much personal information about the astronauts that is not available elsewhere.

_____. *The First Lunar Landing: 20th Anniversary/as Told by the Astronauts, Neil Armstrong, Edwin Aldrin, Michael Collins.* Washington, DC: National Aeronautics and Space Administration EP-73, 1989. This is a short recollection of the Apollo 11 mission by the astronauts.

Atkinson, Joseph D., Jr., and Shafritz, Jay M. *The Real Stuff: A History of the NASA Astronaut Requirements Program.* New York: Praeger Pubs., 1985. The authors present a solid overview of the selection of the NASA astronauts and their development.

Benson, Charles D. and Faherty, William Barnaby. *Moonport: A History of Apollo Launch Facilities and Operations.* Washington, DC: National Aeronautics and Space Administration SP-4204, 1978. An excellent history of the design and construction of the lunar launch facilities at Kennedy Space Center.

Bergaust, Erik. *Murder on Pad 34.* New York: G.P. Putnam's Sons, 1968. A highly-critical account of the investigation of the Apollo 204 accident in January 1967 that killed astronauts Gus Grissom, Roger Chaffee, and Ed White.

Bilstein, Roger E. *Stages to Saturn: A Technological History of the Apollo/Saturn Launch Vehicles*. Washington, DC: National Aeronautics and Space Administration SP-4206, 1980. This thorough and well-written book gives a detailed but highly readable account of the enormously complex process whereby NASA and especially the Marshall Space Flight Center under the direction of Wernher von Braun developed the launch vehicles used in the Apollo program ultimately to send 12 humans to the Moon.

Booker, Peter Jeffrey; Frewer, G.C.; and, Pardoe, G.K.C. *Project Apollo: The Way to the Moon*. New York: American Elsevier Pub. Co., 1969. A popular and readable account prepared in anticipation of and released just after the Apollo 11 mission in 1969, this book condenses the essential details of 10 years of American space activities into a short narrative.

Borman, Frank. *Countdown: An Autobiography*. New York: William Morrow, Silver Arrow Books, 1988. With Robert J. Serling. Written to appear on the twentieth anniversary of the first lunar landing, this autobiography spans much more than the Apollo program.

Breuer, William B. *Race to the Moon: America's Duel with the Soviets*. Westport, CT: Praeger, 1993. This book, written by a journalist who has made a career out of writing World War II adventures, is neither about the race to the Moon, nor the U.S. rivalry with the U.S.S.R. The majority of it is, instead, about the World War II efforts of the German rocket team under Wernher von Braun at Peenemuende, their wartime exploits, their surrender to American forces in 1945, and their post-war activities in the U.S.

* Brooks, Courtney G., Grimwood, James M., and Swenson, Loyd S., Jr. *Chariots for Apollo: A History of Manned Lunar Spacecraft*. Washington, D.C.: National Aeronautics and Space Administration SP-4205, 1979. The authors of this book describe the development of the spacecraft used in Project Apollo.

---

*Available as a Dover reprint.

380

Collins, Michael. *Carrying the Fire: An Astronaut's Journeys*. New York: Farrar, Straus and Giroux, 1974. This is the first candid book about life as an astronaut, written by the member of the Apollo 11 crew that remained in orbit around the Moon. The author comments on other astronauts, describes the seemingly endless preparations for flights to the Moon, and assesses the results.

_____. *Liftoff: The Story of America's Adventure in Space*. New York: Grove Press, 1988. A general history of the U.S. space program for a popular audience written by a former astronaut, begins with an account by one of the three participating astronauts of the Apollo 11 flight. He then flashes back to the post-World War II beginnings of the United States' interest in space and traces the evolution of the space program through the history of the Apollo program.

* Compton, W. David. *Where No Man Has Gone Before: A History of Apollo Lunar Exploration Missions*. Washington, DC: National Aeronautics and Space Administration SP-4214, 1989. This clearly-written account traces the ways in which scientists with interests in the Moon and engineers concerned with landing people on the Earth's satellite resolved their differences of approach and carried out a mission that made major contributions to science and developed remarkable engineering achievements.

Cooper, Henry S.F. *Apollo on the Moon*. New York: Dial Press, 1969. In this book Cooper predicts, before the landing of Apollo 11 astronauts on the Moon in July 1969, what they would encounter. More important, he follows the preparations for the mission with great skill and recounts them in his personal and scintillating style. A small work, this book is barely 140 pages and is taken almost verbatim from two of Cooper's *New Yorker* articles.

---

*Available as a Dover reprint.

_____. *Moon Rocks*. New York: Dial Press, 1970. This is an informal account of the first investigating team's examining the lunar samples at Houston.

_____. *Thirteen: The Flight that Failed*. New York: Dial Press, 1973. A lively account of the nearly-disastrous flight of Apollo 13.

Fries, Sylvia D. *NASA Engineers and the Age of Apollo*. Washington, DC: National Aeronautics and Space Administration SP-4104, 1992. This book is a sociocultural analysis of a selection of engineers at NASA who worked on Project Apollo.

Furniss, Tim, *"One Small Step"—The Apollo Missions, the Astronauts, the Aftermath: A Twenty Year Perspective*. Somerset, England: G.T. Foulis & Co., 1989. Developed as a retrospective celebration on the twentieth anniversary of the lunar landing, this book tries to recreate the exhilaration of the Apollo missions.

Gray, Mike. *Angle of Attack: Harrison Storms and the Race to the Moon*. New York: W.W. Norton and Co., 1992. This is a lively journalistic account of the career of Harrison Storms, president of the Aerospace Division of North American Aviation that built the Apollo capsule.

Hallion, Richard P., and Crouch, Tom D. Editor. *Apollo: Ten Years Since Tranquility Base*. Washington, DC: Smithsonian Institution Press, 1979. This is a collection of essays developed for the National Air and Space Museum, commemorating the tenth anniversary of man's first landing on the Moon, July 20, 1969.

Hoyt, Edwin P. *The Space Dealers: A Hard Look at the Role of Business in the U.S. Space Effort*. New York: The John Day Co., 1971. This book describes the intricate interrelationships between government organizations such as NASA and the aerospace industry. Not specifically focused on Project Apollo, it uses it as a test case in looking at the larger question of government/industry relations.

Kennan, Erlend A., and Harvey, Edmund H., Jr. *Mission to the Moon: A Critical Examination of NASA and the Space Program.* New York: William Morrow and Co., 1969. This book features a detailed examination of the facts of the Apollo 204 fire in January 1967 that killed three astronauts. It does not provide a balanced account of the lunar landing program or NASA. Instead it is filled with critical asides.

Launius, Roger D. *NASA: A History of the U.S. Civil Space Program.* Melbourne, FL: Krieger, 1994. A short book in the Anvil Series, this history of U.S. civilian space efforts consists half of narrative and half of documents. It contains three chapters on the Apollo program, but while coverage consists more of overview than detailed analysis, the approach is broadly analytical and provides the most recent general treatment of its topic.

Lewis, Richard S. *Appointment on the Moon: The Inside Story of America's Space Adventure.* New York: Viking, 1969. Perhaps the first book to capitalize on the success of Apollo 11 in 1969, this history appeared within days of the "splashdown."

Logsdon, John M. *The Decision to Go to the Moon: Project Apollo and the National Interest.* Cambridge, MA: The MIT Press, 1970. This book describes in detail the political issue of how the United States decided to go to the Moon in 1961.

McDougall, Walter A., . . .*The Heavens and the Earth: A Political History of the Space Age.* New York: Basic Books, 1985. This Pulitzer Prize-winning book analyzes the space race to the Moon.

Mailer, Norman. *Of a Fire on the Moon.* Boston: Little, Brown, 1970. London, Weidenfeld & Nicolson, 1970. New York: New American Library, 1971. One of the foremost American writers, Mailer was commissioned to write about the first lunar landing. But he was forced, grudgingly, to admit that NASA's approach to task accomplishment—which he sees as the embodiment of the Protestant Work Ethic—and its technological and scientific capability got results with Apollo.

Mansfield, John M. *Man on the Moon.* New York: Stein and Day, 1969. Written by a BBC television producer, this book begins with ancient conceptions of the Moon and continues with theoretical foundations for the space age in their works of science fiction authors and theoreticians. The book's capstone is a discussion of NASA and Project Apollo.

Masursky, Harold; Colton, G.W.; and El-Baz, Farouk. *Apollo Over the Moon: A View from Orbit.* Washington, DC: National Aeronautics and Space Administration SP-362, 1978. This is an excellent encapsulation of the Apollo program with striking photography.

Murray, Charles A., and Cox, Catherine Bly. *Apollo, the Race to the Moon.* New York: Simon and Schuster, 1989. Perhaps the best general account of the lunar program, this history uses interviews and documents to reconstruct the stories of the people who participated in Apollo.

Rabinowitch, Eugene, and Lewis, Richard S. Editors. *Man on the Moon: The Impact on Science, Technology, and International Cooperation.* New York: Basic Books, 1969. The editors have assembled articles that provide a range of views on the impact of the exploration of space on science, technology, and international cooperation. Each author approaches the subject from their own perspective, speculating on the meaning of the Apollo lunar landing and offering prognostications for the future.

Thomas, Davis. Editor. *Moon: Man's Greatest Adventure.* New York: H.N. Abrams, 1970. A large-format, illustrated work, the centerpiece of this book are three major essays. One, by Fred A. Whipple, Harvard University astronomer, describes the possibilities of space flight for scientific inquiry. Another by Silvio A. Bedini, Smithsonian Institution, deals with the Moon's role in human affairs. A final article by Wernher von Braun, of NASA, analyzes Project Apollo and its execution in the 1960s.

von Braun, Wernher. *First Men to the Moon.* New York: Holt, Rinehart and Winston, 1966. A popular account of Apollo based of a series of articles appearing in *This Week* magazine. Its greatest strength is the inclusion of easily understood diagrams of scientific phenomena and hardware.

Wilford, John Noble. *We Reach the Moon: The New York Times Story of Man's Greatest Adventure.* New York: Bantam Books, 1969. One of the earliest of the journalistic accounts to appear at the time of Apollo 11, a key feature of this general and journeyman but not distinguished history is a 64-page color insert with photographs of the mission. It was prepared by the science writer of the *New York Times* using his past articles.

Young, Hugo, Silcock, Bryan, and Dunn, Peter. *Journey to Tranquillity: The History of Man's Assault on the Moon.* London: Cape, 1969. Garden City, NY: Doubleday, 1970. A ponderous "anti-Apollo" broadside, this book seeks to cast aspersions on the entire space program.

ABOUT THE EDITOR

Glen E. Swanson has held a fascination for space exploration since childhood. A native of Grand Rapids, Michigan, he received a bachelor of science degree in education from Western Michigan University in 1987, where he majored in history and minored in physics and mathematics. After several years of teaching, including a year as program director for the Michigan Space Center in Jackson, Michigan, he founded CSPACE Press, an aerospace publishing firm, during which time he acquired, and served as editor of, *Countdown*, a monthly magazine covering the Space Shuttle program. In 1992, he founded *Quest*, the world's only publication devoted to the history of space flight. Swanson has served as a consultant for several space history television productions, including *Moon Shot*, *Rockets*, and the HBO series *From the Earth to the Moon*. After obtaining a master of science degree in space studies from the University of North Dakota in 1998, he began working for NASA as historian of the Johnson Space Center in Houston, Texas. He currently lives in League City, Texas.

INDEX